Confronting the Political in International Relations

Edited by

Michi Ebata
Research Student
London School of Economics

and

Beverly Neufeld
Research Student
London School of Economics

in association with
MILLENNIUM JOURNAL OF
INTERNATIONAL STUDIES

First published in Great Britain 2000 by
MACMILLAN PRESS LTD
Houndmills, Basingstoke, Hampshire RG21 6XS and London
Companies and representatives throughout the world

A catalogue record for this book is available from the British Library.

ISBN 0-333-73285-5 hardcover
ISBN 0-333-73286-3 paperback

 First published in the United States of America 2000 by
ST. MARTIN'S PRESS, LLC,
Scholarly and Reference Division,
175 Fifth Avenue, New York, N.Y. 10010

ISBN 0-333-73285-5

Library of Congress Cataloging-in-Publication Data
Confronting the political in international relations / edited by Michi Ebata and Beverly Neufeld.
   p. cm.
Includes bibliographical references and index.
ISBN 0-333-73285-5
  1. International relations—Political aspects. I. Ebata, Michi. II. Neufeld, Beverly.

JZ1253 .C66 2000
327.1—dc21
                                                                00-042247

© Millennium 2000

All rights reserved. No reproduction, copy or transmission of this publication may be made without written permission.

No paragraph of this publication may be reproduced, copied or transmitted save with written permission or in accordance with the provisions of the Copyright, Designs and Patents Act 1988, or under the terms of any licence permitting limited copying issued by the Copyright Licensing Agency, 90 Tottenham Court Road, London W1P 0LP.

Any person who does any unauthorised act in relation to this publication may be liable to criminal prosecution and civil claims for damages.

The authors have asserted their rights to be identified as the authors of this work in accordance with the Copyright, Designs and Patents Act 1988.

This book is printed on paper suitable for recycling and made from fully managed and sustained forest sources.

10 9 8 7 6 5 4 3 2 1
09 08 07 06 05 04 03 02 01 00

Printed and bound in Great Britain by
Antony Rowe Ltd, Chippenham, Wiltshire

Contents

List of Contributors	vi
1. Politics in International Relations *Michi Ebata and Beverly Neufeld*	1
2. The Development of the Concept of International Society: An Essay on Political Argument in International Relations Theory *Edward Keene*	17
3. Culture and International Relations: A New Reductionism? *Fred Halliday*	47
4. Global Politics and the Problem of Culture: The Case of China *Christopher Hughes*	72
5. Overturning 'Globalisation': Resisting the Teleological, Reclaiming 'the Political' *L. Amoore, R. Dodgson, B. Gills, P. Langley, D. Marshall & I. Watson*	98
6. Theorising Politics in 'No Man's Land': Feminist Theory and the Fourth Debate *Bice Maiguashca*	123
7. What is Left of the Domestic? A Reverse Angle View of Foreign Policy *Christopher Hill*	151
8. The Politics of Place and Origin: An Inquiry into the Changing Boundaries of Representation, Citizenship, and Legitimacy *Friedrich V. Kratochwil*	185
9. International Relations Theory and the Fate of the Political *R.B.J. Walker*	212
Index	239

Contributors

L. Amoore, R. Dodgson, B. Gills, P. Langley, D. Marshall, and I. Watson are members of the Newcastle Research Working Group on Globalisation, University of Newcastle upon Tyne. Gills is Director of the Programme in International Political Economy at the Department of Politics, University of Newcastle upon Tyne. He has published widely in the fields of international political economy, development and Korean politics.

Michi Ebata organised the *Millennium: Journal of International Studies* 25th Anniversary Conference at the LSE with Stefan Fritz. Her PhD thesis is entitled *The transition from war to peace: politics, political space and the peace process industry in Mozambique, 1992-1995*. She is joining the Department of War Studies at King's College London as a MacArthur Post-Doctoral Research Associate.

Fred Halliday is Professor of International Relations at the LSE. His most recent books are *Rethinking International Relations* (1994) and *Islam and the Myth of Confrontation* (1996).

Christopher Hill is Montague Burton Professor of International Relations at the LSE. His main interests are foreign policy analysis and the external relations of the EU and its Member States. Relevant publications include: Cabinet Decisions on Foreign Policy: the British Experience October 1938-June 1941, jointly edited with Pamela Beshoff *Two Worlds of International Relations: Academics, Practitioners and the Trade in Ideas* and edited *The Actors in Europe's Foreign Policy*.

Christopher Hughes lectures in the international relations of Asia-Pacific, with special reference to China at the University of Birmingham, and will join the Department of International Relations at the LSE. He has published *Taiwan and Chinese Nationalism* and various journal articles on China and the global system.

Edward Keene is lecturer in International Politics at the School of Oriental and African Studies, University of London. He completed a PhD in International Relations at the LSE on the concept of international society and the development of modern international society. He is a former editor of *Millennium: Journal of International Studies* (1995) and is the co-editor of *The Globalisation of Liberalism* (forthcoming).

Friedrich Kratochwil holds the Chair for International Politics at the University of Munich. He studied philosophy and ancient history at Munich University and received a Master's from Georgetown and a PhD in International Relations from Princeton. He is the author of *International Order and Foreign Policy: Rules, Norms and Decisions* as well as several edited and co-authored books. His present research concerns social and legal theories in international politics.

Bice Maiguashca is lecturer in the Department of Politics at Exeter University. She is a former editor of *Millennium: Journal of International Studies*, and is currently researching social movements, social theory and their relevance to contemporary debates in international relations theory.

Beverly Neufeld is a research student in International Relations at the LSE. Her thesis is entitled *Territory, Identity, Politics: The Concept of Community in International Relations*. She has published 'Feminism and the Concept of Community' in S. Chan and J. Weiner (eds.) *Currents of International Relations* (1998), and previously in *Millennium: Journal of International Studies*.

R.B.J. Walker is Professor of Political Science at the University of Victoria, B.C., Canada and Director of the Interdisciplinary Graduate Program in Contemporary Social and Political Thought, and editor of the journal *Alternatives*.

1. Politics in International Relations
Michi Ebata and Beverly Neufeld[1]

> This essay simply seeks to help in the task of restoring confidence in the virtues of politics as a great and civilizing human activity.[2]
>
> Bernard Crick

It seems redundant to suggest that international relations is about politics. The discipline emerged in response to the political problem of war, and from its inception was concerned with explaining and understanding the nature of international politics. With the professionalisation of the discipline, however, it seems that those studying international relations increasingly pay more attention to political issues, and less attention to what constitutes politics and the political in international relations. Thus while it seems unnecessary to state that politics is embedded in every aspect of international relations, it is increasingly important to emphasise the fundamentally political nature of the discipline and to explore the implications of what this means. In considering the political in international relations, this volume focuses attention both on how politics informs the discipline, and on what limits are placed on politics in international relations.

In the introduction, we argue that politics is not only an essentially contested concept, but that contest and debate is in fact the very substance of politics. We begin by looking at three different perspectives on the concept by Max Weber, Bernard Crick and William Connolly. The second section considers the question of how politics is understood in the discipline of international relations, by focusing on the dichotomy of a discipline that has seen the notion of politics greatly expand during this century, and yet still tends to conceive of politics and the political as largely fixed and unchanging. The third section addresses the question of what a reconsideration of politics might offer international relations, arguing that reexamining the meaning of politics in international relations involves emphasising the notion of contest in the discipline, both in terms of what we study and how we study it. In the fourth section we provide brief overviews of the chapters in the volume. Each author addresses the need to engage with politics and the political in international relations, and explores the implications of a reconsideration of the political for the discipline. Finally, we discuss three central themes on politics in international relations raised throughout the volume, and conclude with some thoughts on what a reconsideration of politics by international relations might offer.

Why talk about politics?

There are few obvious parameters for a discussion of politics in international relations, since all that comprises international relations is political. But if international relations is inescapably about politics, then the way in which the concept of politics is defined has important implications for understanding the discipline. Within the study of philosophy and the discipline of political science there is a long history of debate on the meaning of politics. In the literature of international relations, definitions of politics may be implied, but for the most part, international relations has been content to borrow its conceptions of the political from other fields without much debate or deliberation. In this section, therefore, we draw on conceptions of the political from sociology, political science and political philosophy, considering three definitions of politics from Max Weber, Bernard Crick and William Connolly.

One traditional and highly influential understanding of politics is set out in Max Weber's 'Politics as a Vocation'. Weber argues that

> a state is a human community that (successfully) claims the monopoly of the legitimate use of physical force within a given territory... The state is considered the sole source of the "right" to use violence. Hence, "politics" for us means striving to share power or striving to influence the distribution of power, either among states or among groups within a state.[3]

Weber's definition of politics, then, is limited to the boundaries of state power relations, where power is a measurable quality that is possessed. For Weber, '[t]he decisive means for politics [in the modern state] is violence'[4]. But he argues that the morally dubious necessities of politics might be mitigated by political '...conduct that follows the maxim of an ethic of responsibility, in which case one has to give an account of the foreseeable results of one's action.'[5] So for Weber, the question of what it means to act ethically is as important as the question of what it means to act politically.

However, Weber's concern with the ethics of political behaviour is limited by his conception of politics which is linked to a particular definition of the state and state power. Where Weber draws out politics from the power structures of the state, Crick is less interested in the locus of power in the state. Although he assumes politics takes place within the state, he does not focus on the location of politics, but instead on the function of politics in organised societies. Crick conceptualises politics as an activity in itself. Given inevitable differences within society, the activity of politics is necessary for framing debate about these differences, leading to forms of accommodation that prevent society from disintegrating. For Crick, '...politics represents at least some tolerance of differing truth, some recognition that government is possible, indeed best conducted, amid the open canvassing of rival interests.'[6] The activity of politics is an open-ended process of discussion, debate and reconciliation that sustains the cohesion of organised society. If politics is indeed an activity, then as Christopher Hughes argues, '[t]he task of politics ...should not be to eliminate rival ideological positions, but to create the conditions within which discussion can take place between them.'[7]

It would seem, then, that defining politics as either an activity or as about power relations in and between states could be understood as a political act itself. Indeed, for William Connolly, *debating* the meaning of politics is a political activity because politics does not have one settled definition. Drawing on W.B. Gallie, Connolly argues that politics is an essentially contested concept. He suggests that

> [p]olitics is the sphere of the unsettled, [it is] at its best, simultaneously a medium in which unsettled dimensions of the common life find expression and a mode by which a temporary or permanent settlement is sometimes achieved...[8]

Thus, politics is not only an essentially contested concept, but contest is also the very substance of politics. Connolly shares Crick's ideas on politics as a process of debate and discussion, but, for Connolly, politics is not only an activity but also an arena where contending differences are articulated and challenged and which may or may not be reconciled. Connolly argues that '[t]here is no contradiction in first affirming the essential contestability of a concept and then making the strongest case available for one of the positions within that range. That's politics.'[9]

This brief overview of three definitions of politics illustrates the contested nature and diversity of meanings of the political. In the next section, we explore the tension between the contested nature of politics and the international relations conception of politics that seems given and settled. The need for a better understanding of the contested nature of politics in international relations is discussed in the context of an overview of the history of the discipline.[10]

How does international relations understand politics?

The history of the academic discipline of international relations is relatively brief. Beginning from its explicitly political roots in the anti-war movement following the First World War, its history is inseparable from the political and social upheavals that define the twentieth century. As Fred Halliday points out, '[i]nternational relations exists as an academic subject because of, and in permanent tension with, the world of history and events.'[11] In addition, the study of international relations has also been characterised by theoretical and methodological disputes. The dominant paradigm, realism, has been challenged by idealists, pluralists, structuralists, and most recently, poststructuralists. These theoretical and methodological contests have provided frequent opportunities for international relations to review its progress and assess its relevance and its place in the social sciences. As debates which characterise the history of the discipline, they may also be seen as contests about the nature of politics. However, even though some of the most interesting political questions of this century have been, and continue to be, asked within international relations, these contests have not produced an *explicit* debate about politics in the discipline.

The traditional understanding of politics found in international relations under the banner of realism relies on a Weberian notion of politics that is state-centric and power-based. Initially, the realist conception of politics was largely confined to a notion of 'high politics', to a concern with power balances, order and security, survival and anarchy. Politics in international relations was power politics, and, consequently, strict limits were imposed on what was understood to be political. Problems of justice, change or social development and democracy were the responsibilities of states to their own populations. The role of individuals and the internal domestic affairs of states, 'low politics', were mostly ignored, because there was an assumption that politics as power occurred only between states, since power was a commodity belonging only to states.

Increasingly, however, these distinctions of 'high' and 'low' politics proved untenable. It seemed to many working in the field that there was more to the international system than simply the state as actor, with survival as its object. International relations scholars began to attempt to locate politics at the intersections between states, societies and the international. As a consequence, there was an expansion in research interests reflecting a recognition of social and political actors beyond the state. Foreign policy analysis, international institutions and international political economy became legitimate areas of research, as did normative concerns, feminism, Marxism, the environment, and social movements. The discipline began to recognise that politics may be located in arenas beyond and below the state, and undertaken by actors other than states, with objectives not necessarily related to the state.

To some extent, these developments reflect the conception of political activity in the Crick tradition. Crick's definition of politics as an activity allows for an understanding of politics beyond the state that focuses on the interactions amongst different kinds of political actors. Thus it would seem that these theoretical and methodological debates allow for differing conceptions of politics in the discipline, or at the very least for a debate about politics in international relations. But the recognition of more political *issues* in international relations was not accompanied by an examination of the discipline's understanding of politics. Hedley Bull's comments about international politics in 1972 still resonate more than twenty-five years after they were first written. In international relations, he wrote,

> [w]e still encounter writings which, while they display sophistication in the handling of one or another technique for the study of international politics, display innocence about international politics itself. There are reasons for thinking that this kind of innocence is more or less perennial, that the attempt to by-

pass or circumvent problems of a political nature is bound to keep cropping up in one form or another. But in general the explanations now provided of international events are deeper and more many-sided than they were fifty years ago.[12]

Thus, while Bull would like to minimise the problem by claiming increased sophistication and 'deeper and more many-sided' explanations in international relations, the fact remains that it is rare to find an explicit discussion of politics in the discipline. Ironically, then, opening up the discourse of international relations to new concepts and questions has meant that the study of international relations has expanded, but at the same time, its understanding of politics has not. Instead, theoretical and methodological debates have mainly served to clarify the relatively narrow parameters of politics that the fundamentally realist conception of the political established in international relations. As a result, while the plurality of the discipline is increasingly evident, this growth has not opened up the scope for an expanded debate about politics. In fact, international relations is drifting further and further away from its political antecedents.

To take one example, recent developments in international relations seem to suggest that the importance of the state is declining or even absent, and that as a result the state-centric definition of politics is increasingly bankrupt. And yet while the discipline is increasingly looking away from the state, this shift has not produced questions concerning the conception of politics in international relations, but only challenges to the boundaries of the state as a *locus* for politics. This has not been accompanied by a reconsideration of either the *function* or the *processes* of politics, and thus the problem is that in dropping the state, a range of crucial political relations are ignored. As Christopher Hill points out '[political action] and agency may be possible beyond the confines of the state, but there is as yet no clear way to organise domestic accountability without falling back on the formations of an organised polity and of society.'[13] In other words, while there is a recognition in international relations that its conception of politics is inadequate, the discipline has not yet entered into debate about the notion of politics itself.

Perhaps part of the problem in talking about politics in international relations may be that what is political for one generation of scholars may or may not be for the next. Indeed, what some would seek to politicise in the discipline, others would dismiss as having little or nothing to do with politics or with international relations. But by dismissing one conception of politics and replacing it with another, international relations fails to recognise that these different notions of the political represent a contest about the meaning of politics, which as Connolly notes, is the *essence* of politics. The discipline is impoverished by ignoring the opportunity to engage in an explicit debate to contest the notion of politics in international relations.

In the next section, we adopt Connolly's conception of politics as contest to argue that debates about what constitutes politics are crucial for revealing the nature of politics in international relations. We argue that international politics needs to be conceptualised by confronting the contested nature of politics and the political in international relations.

What can a reconsideration of politics offer international relations?

It is clear from the history of the international relations discipline that it now has more thematic scope and theoretical depth than ever before. Indeed, as a relatively young field, a certain amount of disciplinary insecurity has found international relations often engaging in self-assessment, occupying itself with questions of its own relevance, and appearing to make progress in doing so. But in spite of this rich tradition of questioning its *raison d'etre*, more often than not self-justification rather than self-criticism (which might lead to a more sophisticated understanding of the political) has been the result.

Part of the problem lies in the fact that international relations has largely overlooked the core questions of politics and the political, and apparently has yet to recognise that politics is an essentially contested core concept. Perhaps one of the reasons that international relations seems to be in a perpetual state of disciplinary insecurity is that it has never explicitly examined its conception of politics. After all, taking stock of the discipline and its accomplishments differs from critically examining the core assumptions of the field.

One of the key concerns of the discipline today is that the empirical, methodological and theoretical plurality in the discipline both opens it up and threatens to pull it apart. This problem may be overcome by accepting that politics is an open-ended process, an activity. The point is not to achieve a unified perspective on all issues in international relations or to point the discipline in a single direction. Rather, understanding politics as itself contested and open-ended might provide the means for more intra-disciplinary discussion, instead of the isolated debates taking place today.

For example, Bice Maiguashca argues that poststructuralists and critical theorists are constantly (and somewhat ironically) talking past each other, and that

> the discipline of International Relations remains largely devoid of any conceptual language that would enable us to identify and talk about social oppression, social resistance and social movements. Without such conceptual tools, I believe that it is simply impossible to speak about contemporary politics in our discipline.[14]

Thus the post-positivists are talking mainly (and somewhat unsuccessfully) amongst themselves. And while the more traditional in the discipline are still engaged in either challenging realism or defending it, there seem to be fewer debates and more overlapping monologues. A focus on the question of politics might provide both a language for debate amongst different approaches in international relations, and also an opportunity to engage the political by reexamining our understanding of political argument.

However, ignoring politics has greater implications than simply intradisciplinary communication. An uncontested understanding of politics may lead to political *processes* that either lack meaning or create a disjuncture with reality. Friedrich Kratochwil explains that

> [t]here seems to be a growing disparity between the practices which comprise international relations, and the conceptual apparatus by which we attempt to analyse these practices... Equally surprising, though, is the fact that the treatment of long-term secular changes seems to have little impact on the conceptual lenses by which we analyse the ongoing practices of international politics.[15]

The problem is that the international relations discipline appears to have forgotten that the academic who studies international politics is engaged in a political act and inevitably takes on a measure of political responsibility. Paraphrasing William Connolly, time spent examining the concepts of politics is *not* time taken away from politics itself. Rather, '... conceptual debates... are so often intense because we tacitly understand the relation of these debates to our deepest commitments and we sense as well the import that the outcome of such contests has for the politics of our society.'[16]

Thus, what is at stake in international relations debates is the very meaning of politics. We contend that it is a fundamental problem for international relations to ignore questions of the political, not only because it can potentially render the discipline irrelevant, and not only because it does make intra-disciplinary communication that much more difficult, but also because it is politics that provides the discipline with its focus and which makes it pertinent.

Confronting the political in international relations

All of the chapters in this volume confront the political in international relations. While not providing an exhaustive account of the discipline, the chapters do challenge the traditional ways in which politics is understood throughout international relations, and the authors set out a number of interesting and important ways in which international relations might reconsider politics.

In Chapter One, Edward Keene traces the evolution of a core concept of international relations, the concept of international society. In *The Development of the Concept of International Society: An Essay on Political Argument in International Relations Theory*, Keene makes the case that international society was once an essentially contested concept and the subject of political argument. In the history of the discipline, this particular argument was won by Hedley Bull, closing off subsequent debate about the concept of international society. The thrust of Keene's argument is that 'the change in the use of the concept of international society illustrates the death of a tradition of argument in international relations and the end of agonism in this particular field of enquiry into world politics.'[17] Keene uses the idea of agonistic political argument to make the point that debates about the proper use of the concept of international society are debates about the kind of politics that emerge from international society. He argues that disputes about political concepts are more than mere debates about conceptual language and that they in fact constitute arguments about our understanding of politics.

Fred Halliday picks up on a number of Keene's concerns about political argument in Chapter Two, *Culture and International Relations: A New Reductionism?* Halliday questions the basis for the resurgence of the study of culture in international relations because of the uncritical adoption of the concept by those who study it. Halliday insists that international relations needs to interrogate and problematise current conceptions of culture, arguing that it must be seen as inherently political. He sets out a broad research agenda for the study of culture by focussing on its political content which opens it to debate and contest, since the influence of culture is 'as much what use existing states and contenders for state power are making of culture, as that of how culture is an independent influence on politics, or how it is weakening the power of states.'[18] Drawing from the work of Antonio Gramsci, Halliday stresses that employing the concept of culture in international relations depends on recognising that 'culture is not an alternative to concepts of economic and political power but a constituent part of their reproduction.'[19]

In Chapter Three, *Global Politics and the Problem of Culture: The Case of China*, Christopher Hughes considers culture in a different sense. He applies the concept of culture to political argument, using the example of how China has engaged in political arguments about culture in developing its own understanding of international politics. Hughes believes that international relations has failed to account for politics as an activity that takes place between and across cultures. Rather than accepting that cultural difference leads to either a homogenous world culture or a heterogeneity of irreconcilable cultures, Hughes draws upon Crick's notion of politics as an activity that enables the reconciliation of differences. He looks at

> how the transmigration of political ideas amounts to far more than merely reproducing or rejecting alien ideas. It is a matter of indigenisation through reinterpretation, a process that results in what has been called "hybrid" concepts... In global politics, hybrid concepts can play a key role in the development of new meanings.[20]

Hughes argues that if such cross-cultural political arguments are to occur more frequently and with less disruption, then a central task for international relations is to explore the conditions under which global issues can make sense in local terms through a process of conceptual transaction and discussion between cultures.

Where Hughes suggests that global politics might be located in a potential dialogue between cultures, Barry Gills et. al., address the idea of globalisation as a political process with social implications. In Chapter Four, *Overturning 'Globalisation': Resisting the Teleological, Reclaiming 'the Political'*, the authors critique the concept of globalisation, arguing that it is not an inexorable process, and that in fact globalisation is a contested concept. For these authors, to posit the process of globalisation as inevitable is to deny politics and to ignore social agency. They suggest that '[s]ocial agency in the form of state, social movements, and organised labour are all basic to forms of politics we currently understand and practice, yet the teleology of globalisation tends to produce the idea of the "death of politics" as well as the demise of the nation-state.'[21] Gills, et. al. argue that highlighting social agency opens up space for a politics of resistance from within the framework of the state, which is local but which takes on global issues, and that in turn this allows for an expansion of the political in international relations.

The theme of social and political resistance is also addressed by Bice Maiguashca in Chapter Five, *Theorising Politics in 'No Man's Land': Feminist Theory and the Fourth Debate*. Like Gills et. al., Maiguashca contends that politics is located in the social sphere, and she argues that as a social movement, feminism has the potential to offer a new conceptualisation of politics in international relations. In particular, she suggests that feminist theory, grounded in the concrete political practices of a social movement, is able to offer a way of moving past the impasse of the 'fourth debate' in international relations between postmodernism and critical theory. Maiguashca argues that both postmodernism and critical theory address two of the most pressing issues of contemporary international relations: the 'politics of redistribution' and the 'politics of recognition'. In making her case for the importance of feminist theory in post-positivist international relations, Maiguashca also reminds intellectuals of their responsibility for political practice, noting that '...one of the main aspirations of post-positivist international relations theorists is to construct a theory of politics that is more sensitive to the various forms of oppression and resistance in contemporary international life.'[22]

Christopher Hill continues the project of looking into society to find ways of responding to world politics. In Chapter Six, *What is Left of the Domestic? A Reverse Angle View of Foreign Policy*, Hill argues that challenges to realism have resulted in the dismissal of foreign policy as a useful tool of analysis. Hill challenges the trend that blurs the boundaries between foreign and domestic politics by seeking to reclaim domestic society as a location for politics. Defining politics as the conflict over the authoritative allocation of values, he argues that politics still takes place primarily in the state. Hill asserts that '[t]he ineluctable tendency of those who wish to relegate the state to a position of lesser importance in international relations, is that of downgrading politics.'[23] While not discounting the insights of some of the new trends in international relations, Hill contends that '[t]he relationship between the domestic and the foreign is always in flux; but it is still crucial to our understanding of politics, that is, what has been done, what can be done, and what should be done.'[24] Without accounting for the state, foreign policy and domestic politics, Hill argues, international relations risks losing a sense of the politics that comprise international relations.

In Chapter Seven, *The Politics of Place and Origin: An Inquiry into the Changing Boundaries of Representation, Citizenship and Legitimacy,* Friedrich Kratochwil examines a number of themes that Hill raises concerning the connections between domestic and international politics. Kratochwil argues that to understand politics, it is necessary to understand what constitutes the state on the one hand and the subjects who give their consent to be governed by the state on the other. Kratochwil argues that 'the dynamics of politics emerges from the definition of the group, the efforts of locating it in a place and in establishing representative institutions entitled to make public choices.'[25] The problem, he notes, is that while the conceptual tools of membership, legitimacy and representation are particular and tied to a place, and have a limited capacity for explanation, international relations nevertheless tends to adopt these tools as universally applicable concepts. Kratochwil suggests that this overlooks the political dynamics of defining the group and locating it in a certain place. Without a fixed referent to people or to place, the kind of politics that emerges is partial and limited and, indeed, the whole project of 'politics seems thoroughly partial.'[26]

Where Kratochwil is concerned with the limits of the conceptual tools with which we understand politics in international relations, R.B.J. Walker focusses on the notion of sovereignty and the limitations it places on modern politics. In Chapter Eight, *International Relations and the Fate of the Political*, Walker suggests that international relations has the opportunity to challenge received notions of politics, and suggests that although important, the new critical literature risks marginalising questions about politics as much as the traditions it challenges. The reason for this risk is '...the uncertain status of the sovereign state as the primary constitutive principle governing what and where modern politics is supposed to be.'[27] Walker argues that state sovereignty and sovereign subjectivity no longer adequately resolve the dichotomy between universality and particularity. At the same time, he suggests that:

> Debates about international relations theory are, however, significantly privileged in this context both because they, perhaps more than any other arena of contemporary thought, are always supposed to be aware both of the *limits* of a modern politics expressed by the principle of state sovereignty and of the consequent contradiction inherent in *any* claim to a politics of the "world", that is, a politics that exceeds the legitimate authority of sovereign territorial jurisdictions.[28]

Walker's concerns with sovereignty and sovereign subjectivity are central to confronting the contested nature of the political. In the next section, we examine the implications of this problem and several other themes raised by the authors for a reconsideration of politics.

Conclusion: What can a reconsideration of politics by international relations offer?

These essays by no means provide an exhaustive account of politics in international relations. However, each seeks to examine both the ways in which politics informs the discipline, and the ways in which limits are placed on the political in international relations. The authors demonstrate that international relations not only has much to gain from a reconsideration of politics, but that there is also much that international relations has to offer.

All of the authors in this volume confront the question of politics in international relations from different vantage points, but several common themes arise throughout their work. One theme is the problem of limiting politics in international relations to the state and its power relations. All of the authors challenge this state-centric definition of political space, not least because in order to talk about politics they frequently cross the traditional boundaries of the state. To the extent that there is a general recognition in international relations that state-based politics no longer captures the variety and scope of international politics, the essays here might appear to simply echo changes already taking place in the discipline. But as we argued earlier, the recognition of more political *issues* does not mean that international relations has re-examined what is understood to constitute politics.

The point is that it is simply not enough to widen the scope of politics by shifting the political boundaries of international relations. To challenge the dichotomy between the international and the domestic is not to suggest that we live in a global village, but rather that the intersections between the international, domestic and local open up new spaces for politics. If, as the authors suggest, new spaces are opened for politics by social movements, the transmigration of ideas between cultures, and how the domestic is constituted, then these emerging political practices serve new functions which have not yet been explored. By problematising the international and re-drawing the boundaries of politics, a platform is established from which to reconsider the functions and the process of politics.

All of the authors outline different ways of taking the next step from this platform. What they share is a concern with the disjuncture that exists between the dynamic practices of international politics and the static conceptual tools of international relations, in addition to the limitations this disjuncture imposes on political argument in the discipline. The authors demonstrate that without examining the discipline's conceptual tools and methods of political argument, the limitations imposed by those redrawn boundaries remain, and politics will continue to inform the discipline only insofar as state power relations appear to allow it. In other words, opening up the scope of the political in international relations could be, and ought to be, accompanied by concurrent developments in the conceptual tools of political argument in international relations. Developing new conceptual tools generates new methods of political argument which in turn multiplies and expands the kinds of contests that can take place about the meaning and practice of politics.

The international relations conception of politics is no longer dependent upon the state, and we increasingly allow for politics to be located elsewhere. By going beyond a mere challenge to the boundaries of political space, then, and questioning the language and methods of political argument in international relations, this volume seeks to take the next step in confronting the political in international relations. What is required is a further confrontation with the political, to develop the conceptual tools and the political language which will help to clarify what we mean by, and understand about international politics. Taking these steps will allow scholars working in international relations to confront politics with greater clarity and coherence, and without necessarily having to look outside the discipline. Therefore, international relations needs to acknowledge and fully explore its fundamentally political nature, for doing so will enhance understanding of what constitutes politics and the political, and a more coherent and comprehensive understanding of international politics will result.

NOTES

1. We would like to thank Mark Hoffman and Michael Banks for their helpful comments on an earlier version of this chapter.
2. Bernard Crick, *In Defence of Politics* (London: Penguin, 1964), p. 15.
3. Max Weber, 'Politics as a Vocation' in H.H. Gerth and C.W. Mills (eds.), *From Max Weber: Essays in Sociology* (London: Routledge, 1948), p. 78, emphasis omitted.
4. *Ibid.*, p.121.
5. *Ibid.*, p.118.
6. Bernard Crick, *In Defence of Politics*, Fourth Edition (London: Weidenfeld and Nicholson, 1962), p.18.
7. Christopher Hughes, in this volume, p. 73.
8. William Connolly, *The Terms of Political Discourse*, Second Edition (Princeton, NJ: Princeton University Press, 1983), p. 227.
9. *Ibid.*
10. See Hedley Bull, 'The Theory of International Politics, 1919–1969' in Brian Porter (ed.), *The Aberystwyth Papers: International Politics 1919–1969* (London: Oxford University Press, 1972), pp. 30–55; Michael Banks, 'The Inter-Paradigm Debate,' in M. Light and A.J.R. Groom (eds.), *International Relations: A Handbook of Current Theory* (London: Pinter, 1985), pp. 7–26; Fred Halliday, 'The Pertinence of International Relations,' in *Political Studies* (No. 38, 1990), pp. 502–16, and 'International relations and its discontents,' in *International Affairs* (Vol. 71, No. 4, 1995), pp. 733–46.
11. Halliday, *ibid.*, 'International relations and its discontents', p. 746.
12. Bull, *op. cit.*, p.51.
13. Christopher Hill, in this volume, p. 163.
14. Bice Maiguashca, in this volume, p. 142.
15. Friedrich Kratochwil, in this volume, p. 185.
16. Connolly, *op. cit.*, p. 40.
17. Edward Keene, in this volume, p. 23.
18. Fred Halliday, in this volume, p. 58.
19. *Ibid.*, p. 67.
20. Hughes, *op. cit.*, p. 81.
21. Barry Gills *et. al.*, in this volume, p. 103.
22. Maiguashca, *op. cit.*, p. 144.
23. Hill, *op. cit.*, p. 156.
24. *Ibid.*, p. 178.
25. Kratochwil, *op. cit.*, p. 187.
26. *Ibid.*, p. 208.
27. R.B.J. Walker, in this volume, p. 212.
28. *Ibid.*, p. 215.

2. The Development of the Concept of International Society: An Essay on Political Argument in International Relations Theory[1]

Edward Keene

Politics is simultaneously contrary and accommodating, disintegrative and consensual. It involves continually questioning social practices or institutions, but it also involves the search for an agreement between people as to how they can collectively organise their lives. As William Connolly puts it, 'Politics is, at its best, simultaneously a medium in which unsettled dimensions of the common life find expression and a mode by which a temporary or permanent settlement is sometimes achieved.'[2] To meet the difficult challenge of reconciling these two functions, Chantal Mouffe suggests that the social order needs to be conceived in terms of 'agonistic pluralism': that is to say, an order that 'is based on a distinction between "enemy" and "adversary". It requires that... the opponent should be considered not as an enemy to be destroyed, but as an adversary whose existence is legitimate and must be tolerated.'[3] One important consequence of this conception of politics is that debates between academic political theorists are just as much part of politics as are debates between anyone else. Indeed, to get the widest possible sense of the issues at stake in a political argument, one has to pay attention to how disagreements between political theorists about the meanings of concepts have been carried on.

This chapter will explore changes in the conduct of one political argument in international relations theory: the dispute about the meaning of the concept of international society. Among contemporary international relations theorists it is widely agreed that the concept of international society refers to the existence of a society of states, which lacks a central governing authority and is founded instead on the normative principle of state sovereignty, the legal rule of nonintervention and the institutions of positive international law and balance of power diplomacy.[4] The concept is generally associated with theorists like Hedley Bull and Martin Wight, who have recently attracted praise for having 'carried the torch for normative international relations theory during the long positivist-realist phase of American International Relations' and who apparently 'stood four-square in the constructivist camp.'[5] However, students of international relations often question the utility of Bull's and Wight's concept of international society to describe a world in which non-state actors play an increasingly important role, in which rule-making is becoming more centralised, and in which the principle of state sovereignty is being compromised or modified by the emerging international legal and moral personality of individuals.[6] This problem has led some international relations theorists to propose alternative concepts, like global society or global civil society, as better ways to make sense of contemporary international phenomena.[7]

On the one hand, then, the tradition of international society theory apparently embodied a methodologically useful approach to the theorisation of world politics; yet, at the same time, the substantive theories of world politics that emerged from that approach were centred on a concept that seems to lack heuristic or analytical power with regard to contemporary international phenomena. Ole Waever neatly summarises the problem:

> Wight, Bull and others succeeded in defining an extremely interesting locale in the International Relations landscape, a place from where one had a view better than most other theories. It was possible to ask excellent questions and combine traditions and theories normally not able to relate to each other. And yet, the work springing from this place has a tendency to become dogmatic and self-assured in its belief that knowing how central the states system is as a means for order, it can ultimately defy all postulates about radical change in the system.[8]

Similarly, R.B.J. Walker recognises that Bull and Wight were members of 'a school of thought that recognises that the categories of international relations theory express some quite profound philosophical and political problems.'[9] However, at the same time '[t]he modern statist resolution of all claims to identity and difference in space and time once again proved quite irresistible.'[10] In short, then, traditional theories about international society appear to be simultaneously fruitful and constraining, stimulating and dogmatic. To understand how this strange anomaly has arisen, we need to look more closely at the terms under which traditional theorists argued about the meaning of the concept of international society.

The argument of the chapter is that the concept of international society should be regarded as an essentially contested concept, in the sense that a number of different interpretations of the element of society in world politics can reasonably be defended. However, at present, students of international relations do not properly appreciate this feature of the concept because of the dominance of Hedley Bull's particular conception of international society. Consequently, a change has taken place in the character of political argument in international relations theory, such that theoretical disputes about the element of society in world politics are now conducted in a critical rather than agonistic idiom, between enemies rather than adversaries. First, the idea that some concepts are essentially contested will be reviewed. The second section will present the case for treating the concept of international society as contested, and the third section will explain how the concept has come to be associated with one exclusive formulation in terms of the society of sovereign states. Finally, the chapter will conclude by discussing the consequences for international relations theory of this change in the use of the concept.

ESSENTIALLY CONTESTED CONCEPTS: AN AGONISTIC MODEL OF POLITICAL ARGUMENT

First, we need to construct some guidelines for analysing the way concepts are used in international relations theory. Let us begin with a brief look at how philosophers have dealt with disagreements about the meaning of concepts. Philosophers, as well as theorists of politics, society or jurisprudence, acknowledge that many of the concepts they use are subject to intractable disputes. They find it hard to give any generally agreeable answers to questions like 'what is art?', 'what is democracy?', or 'what is a right?' This state of affairs has led some philosophers to describe these concepts as 'essentially contested', in the sense that they are always open to different reasonable interpretations.[11] Furthermore, it is argued that these conceptual disputes are not always signs of error or confusion, but might be an unavoidable, and even desirable, feature of academic enquiry. The continuing existence of disagreement about the 'real meaning' of such concepts, 'far from being a cause of philosophical scandal, is rather a proof of the continuing need of philosophy and of vital, agonistic philosophy.'[12]

W.B. Gallie gives a good explanation of how concepts come to be open to different reasonable interpretations. First, he observes that such a concept must be appraisive, in the sense that 'it signifies or accredits some kind of valued achievement.'[13] The idea of appraisive concepts has been most thoroughly developed by William Connolly, who maintains that some concepts 'characterise arrangements and actions from a normative angle of vision.'[14] In such cases, the point of using the concept is bound up with its connection to moral points of view. Nor can we reconstruct the concept in a non-normative way, since 'a concept so cleansed would lay idle... With no point or purpose to serve, the concept itself would fall into disuse.'[15] However, in addition to being appraisive, an essentially contested concept must also depict a particular kind of valued achievement: one that is internally complex, variously describable and open. In other words, '[a]ny explanation of its worth must... include reference to the respective contribution of its various parts or features; yet... there is nothing absurd or contradictory in any of a number of possible rival descriptions of its total worth.'[16] Andrew Mason summarises this point rather well: political theorists 'disagree over how to *describe* values such as freedom, social justice, and democracy properly (and therefore over what counts as freedom, social justice, and democracy) because they can reasonably disagree over how much weight to attach to the elements that go to make up the achievement accredited by concepts such as these.'[17]

Beyond these four conditions of appraisiveness, internal complexity, variable describability and openness, Gallie also maintains that for a concept to be essentially contested, it must be used strategically and there must be a shared 'original exemplar', to which rival uses of the concept approximate.[18] These further conditions are especially important if a disagreement is to be evidence of a contest, rather than simply being a case of confusion since, as Steven Lukes points out, 'essentially contested concepts must have a common core; otherwise how could we justifiably claim that the contests were about the same concept?'[19] Furthermore, the various users of the concept must recognise the relevance of other interpretations than their own;

otherwise they could not be said to be engaged in a contest at all, but would simply be talking past one another. Thus, while the combination of appraisiveness and complexity shows why questions about the proper use of concepts come to be involved with deep-seated philosophical and political disputes, the idea of a 'common core' or 'original exemplar' explains when such a dispute is a conceptual contest and not the result of a confused or vague misuse of a concept. Together, these two aspects of the overall argument provide us with grounds for identifying concepts which are properly contested, in such a way that the contest will not be resolvable through decisive rational arguments for one particular interpretation, nor through a more precise use of the concept in question.

It is important to be able to make distinctions between contested and confused concepts with reasonable authority, because they carry considerable implications for the way in which we should regard attempts to clarify or define the meaning of obscure concepts. In a nutshell, if a concept is simply confused or vague, we should welcome clarification through 'reconstruction' or better definition according to abstract, general principles.[20] On the other hand, if a concept is contested, we should regard an attempted definition as no more nor less than an argument for a particular interpretation and explore its relationship to other reasonable interpretations of the concept within the context of the historical development of an area of academic enquiry. As the legal philosopher H.L.A. Hart puts it, in such circumstances, 'though theory is to be welcomed, the growth of theory on the back of definition is not.'[21]

Overall, then, the thesis that some concepts are essentially contested points towards an *agonistic* approach to political argument. In an agonistic political argument, moral and political disputes are carried on through arguments about the proper way of using particular concepts, rather than through arguments about which concept to use to describe the world. Such arguments are unlikely to lead to a decisive conclusion, but they will not be a free-for-all, since they operate within the broadly defined, widely agreed and flexible parameters established by the concept's 'original exemplar' or 'common core'. If conducted properly, they will tend to be reciprocal and tolerant, gradually approximating towards a better shared understanding of the meaning of the core concept itself. This agonistic approach to political argument is rather similar to Alasdair

MacIntyre's account of a 'living tradition', as one which 'is an historically extended, socially embodied argument, and an argument precisely in part about the goods which constitute that tradition.'[22] To engage in an agonistic political argument is precisely to acknowledge one's embeddedness in the 'continuities of conflict'[23] that constitute such a living tradition of thought and practice, and to seek to sustain the tradition by preserving and enriching, not resolving, that internal conflict. This feeds the view of politics outlined at the beginning of the chapter, as involving the simultaneous unsettling and settling of ways of life.

Of course, the agonistic approach is not the only possible way of conducting a political argument, and it has some important negative features. Most obviously, the way in which a tradition of argument is constituted is in itself political, and not everyone is or can be part of a tradition. There is, if you like, a degree to which the openness of an essentially contested concept is limited, superficial, and maybe even inconsistent. As David Miller points out, agonistic theorists make political judgements when they identify the permissible uses of a shared concept, but then deny the efficacy of political judgements in deciding between different uses of the concept. This seems to suggest either the operation of a double standard or an abdication of responsibility: '[h]aving set foot in the political arena to arrive at the list of permissible uses of a term, why is the philosopher reluctant to take the second step and argue for the correct use?'[24] Alternatively, if the virtue of an essentially contested concept is its openness to different points of view, and its capacity to accommodate irreconcilable value pluralism in political argument, the political decision about what goes into the common core surely rules certain perspectives out of the argument altogether. Thus, although the death of a tradition may be immensely damaging to political argument in some respects, it may be liberating in others, in the sense that the specific character of the tradition may have stifled argument among some people, or silenced some dissidents. However, whether or not a particular tradition is positive or harmful, it is very important that, at the very least, theorists of international relations understand the traditions in which they operate and understand how changes have taken place in the idiom of political discussion within which disputes between international relations theorists are carried on.

In this spirit, let us now turn to look at the development of the concept of international society, from its varied usage in early theories of international law to its celebrated formulation as an 'anarchical society' of sovereign states. The thrust of the following discussion is that the change in the use of the concept of international society illustrates the death of a tradition of argument in international relations and the end of agonism in this particular field of enquiry into world politics.

INTERNATIONAL SOCIETY AS AN ESSENTIALLY CONTESTED CONCEPT

There used to be a long-standing disagreement about the real meaning of international society, which was a genuine and essential conceptual contest rather than a sign that some users of the concept were either mistaken or confused. In accordance with the account of essentially contested concepts given above, this contention requires that disagreements about the concept of international society were possible because that concept is appraisive and complex. Furthermore, there must also have been a 'common core' or an 'original exemplar' of international society, to which all of these different interpretations referred, and which provided a shared foundation upon which their mutual recognition and strategic interaction was established. First, the evidence for the disputed nature of the concept of international society will be presented, then the concept's common core will be identified, and finally its appraisive and complex character will be explained.

Disputes about the concept of international society

At first glance, there seem to be good grounds for supposing that the concept of international society is likely to be a contested one. As Wight observes, '[m]en have been arguing for centuries about what the state really is, what its purpose is, how it holds together and why it should hold together; there is far more scope for argument about the nature of the much more shadowy and insubstantial entity called international society.'[25] Wight himself analyses this disagreement in terms of three different traditions in 'international theory': Realism, Rationalism and Revolutionism. Each tradition has a distinctive broad answer to the question of the meaning of international society, and this difference of opinion is perfectly illustrated by their various ideas about the membership of international society. For Realists, 'it is states that are the subjects of international law, and states alone are the units of the international juridical community.'[26] For Rationalists, 'both states and individuals are capable of being members,'[27] while for Revolutionists 'the only true international society is one of individuals.'[28]

When one goes on to examine the three traditions' ideas about the basis of international legitimacy, it becomes even clearer that the meanings they attach to the concept of international society differ from one to the other. For Realists, the only possible source for binding obligations in international society is the will of the member states themselves.[29] This leads Realists to see the norms and rules that govern behaviour in international society merely as expressions of states' self-interest. While Rationalists recognise the importance of state volition, they prefer to regard international society as also governed by the law of nature (manifest in God's will or in reason). Thus, the Rationalist argument 'offers a different answer to the question about international society. Given this view, international society is a true society, but institutionally deficient; lacking a common superior or judiciary.'[30] Revolutionists see the essence of international society neither in the will of states, nor in the Rationalist version of a Lockean social contract. Instead, the basic Revolutionist claim is that 'international society is none other than the community of mankind.'[31] To be sure, this description is highly aspirational; '"is" here means "is essentially", "ought to be" or "is destined to be"'.[32] On this view, the mistake of the Rationalists is to suppose that there is no more to international society than the phenomenal appearance of an 'institutionally-deficient quasi-society', and to ignore the deeper bonds between human beings that really constitute the essence of international society.

Wight's taxonomy of the three traditions might seem to be rather a flimsy foundation upon which to build an authoritative account of different uses of the concept of 'international society'. For example, it might be argued that Wight's scheme lacks weight since it was only ever a device for presenting the relevant material in the form of lectures; Wight did not prepare it for publication and it cannot be relied on as a fully worked-out treatment of the literature.[33] Against this objection, however, one could reply that, although Wight presents his scheme as a classification of international political thought, when he talks about these three different conceptions of international society he is really just using new labels to describe three very old and well-established branches of international *legal* thinking. The Realist, Revolutionist and Rationalist conceptions of international society correspond to the classic tripartite distinction between positivist, naturalist and 'eclectic' (or 'Grotian') theories of international law.[34]

The interdependence of theories of international law and society is hardly surprising when we remember that most accounts of international law begin from the maxim *'ubi societas ibi jus est'*: '[w]ithout society no law, without law no society.'[35] This maxim indicates that any theory of the sources, content and enforcement of international law necessarily requires a prior conception of international society.[36] Clearly, however, this leaves a great deal of scope for different accounts of what it is about society in world politics that makes it possible to speak of the existence of law. Furthermore, the existence of at least three different accounts of where international law comes from, what its guiding principles are, and how it actually affects international conduct can be taken as compelling evidence that there is a protracted disagreement about the concept of international society. Wight's contribution was to show how these different conceptions of international society sprang from, and contributed to, a disagreement between advocates of more general perspectives on international politics.[37]

The common core of the concept of international society

Even if it is agreed that different conceptions of international society do exist in international legal thinking, there are still two possible explanations for this phenomenon: either it could be a genuine conceptual contest about the proper use of the concept, or it could be that some people are using the concept of 'international society' incorrectly. Perhaps legal positivists, eclectics and naturalists are talking about different concepts, for which they are sloppily using the same word, thus creating conceptual confusion. If this latter possibility is to be ruled out, there must be some common core concept of international society to which all of these different uses refer.

One can construct a common core concept by negatively comparing the concept of international society with an alternative concept. Gallie himself uses this kind of procedure when trying to ascertain the original exemplar for the concept of art.[38] One option here is suggested by Chris Brown's summary of the distinction between the concepts of international society and international system: '[a]n international society differs from an international system because it is a form of order which is normatively based, and not simply the outcome of an interplay of forces.'[39] What the concept of international society does, in other words, is to depict conduct in world politics as rule-governed and meaningful. It should be noted that the concept of international society is also often negatively compared with the concept of international community, and this distinction is typically used to highlight the statist character of the concept of international society. Now, Brown notes that the statism apparently inherent in the concept of 'international society' 'does not preclude reference by theorists of international society to non-state actors.'[40] However, he does not acknowledge that the question of the membership of international society is central to the different ways in which the concept is used, nor does he explore the different sources of obligation that different theorists identify in international society.

As we saw earlier, it is on precisely these points that the various conceptions of international society are distinguished. In several of its conceptions, international society is explicitly taken to mean something other than the idea of a society of sovereign states: either it involves individuals and non-state institutions as well as states; or the appearance of a society of states is a fiction, concealing the real essence of an international society composed exclusively of individuals. In effect, Brown's characterisation of the distinction between international society and international community commits him to accepting one conception of international society as its proper definition. However, this raises the question of the arguments that can be used in support of the claim that this conception is the *proper* use of the concept of international society.

The valued achievement expressed by international society

Let us turn to the other features of an essentially contested concept, and here we can begin by recalling that a concept is appraisive if it 'describes from the moral point of view', and as Connolly noted, if the normative element is taken out of such concepts, there will be little point in using them. The use of the concept of international society does indeed seem to be contingent upon the need to describe international relations in normative terms. For example, James Mayall observes that the concept seems to have 'fallen into disuse' at a time '[w]hen so much theorising has been concerned to avoid explicit argument of a normative or moral kind.'[41] Further evidence for the broadly appraisive character of 'international society' can be found in Alan James's argument that, while it lacks some of the strength of the idea of 'community', the idea of society nevertheless has a certain warmth and intimacy, which he contrasts favourably with the 'rather chilly, distant, and mechanical resonance' of the concept of 'international system.'[42] Admittedly, talk of the appraisiveness of a concept at such a general level will always seem a little vague. The appraisiveness of international society is, however, more clearly confirmed when we ask what reasons people give for using the concept of international society. Then, it can be seen that people typically do use the concept of international society in order to describe a valued achievement.

One reason for valuing international society is that it is seen to provide order in an otherwise anarchic world. This is perhaps the most widely celebrated aspect of international society, in the sense that it makes order in world politics 'the achievement of international society.'[43] When we unpack this idea of international order even as it is used only within Bull's conception of international society, it is easy to see that the order supposedly achieved by international society consists of many different components: the 'preservation of the society of states itself', the protection of state autonomy, peace, the limitation of violence, keeping promises and stability of possession.[44] This is not an exhaustive list, but it clearly indicates that, even if international society is valued only because it represents the achievement of order between states, one can still identify a number of different reasons why it is so valued. In other words, international society, even on this limited formulation, is both appraisive and complex.

If we broaden our discussion to embrace more aspects of Wight's account of disputes about the concept, we can see how other components of international society could be said to constitute part of its valued achievement. One such idea would be the propagation of a 'Western value' of liberal tolerance.[45] Alternatively, one could plausibly claim that 'Western values' do not only concern toleration; rather, they are primarily concerned with 'the development and organization of liberty, especially in the form of... constitutional government.'[46] Notably, in this view, instances of intervention are quite consistent with international society, either because 'the independence and separateness of states is less important than the homogeneity of international society', or because 'of the duty of fellow-feeling and cooperation.'[47] This last claim begins to lead towards a slightly different reason for valuing international society, which would be emphasised within a cosmopolitan Revolutionist conception: the achievement of 'international society' in furthering the realisation of justice and human dignity.

Although this has not been a comprehensive discussion of possible reasons for valuing the achievement represented by international society, it is sufficient to demonstrate that the concept of international society is both appraisive and internally complex. Thus, to sum up the argument of this section, there exist a number of different interpretations of the concept of international society. This disagreement can be regarded as evidence that the concept is essentially contested, because it concerns a common core (norm-governed behaviour) that describes an achievement that can potentially be valued for many different reasons.

HEDLEY BULL'S DEFINITION OF THE CONCEPT OF INTERNATIONAL SOCIETY

It is obvious that something rather dramatic has happened to alter the way in which theorists of international relations use the concept of international society, to the extent that today there seems to be quite widespread agreement that the concept refers exclusively to the idea of an 'anarchical society' of sovereign states upholding a minimal order of co-existence. The key moment of change in this respect can be traced to Bull's account of the concept of international society, and the way in which his arguments attained a dominant position within the broader research programme of the 'English school'.[48] Bull's most momentous innovation was to provide a framework for arguing about the concept of international society, which offered only an abbreviated picture of the contested character of the concept and systematically privileged one particular conception of international society over others. In addition, Bull's argument was advanced within the context of the English school, and although some members of that school resisted his formulation of the concept, they were unable to offer a really effective response because they were working within a relatively narrow set of historiographical assumptions about modern international relations that gave very little support for their preferred alternative conceptions of international society.

Bull's conception of international society

At the outset, it should be acknowledged that Bull was a perceptive and subtle theorist of world politics, and especially of international society. His account of the concept of international society extends over many different works, includes many nuances and also developed over the course of his intellectual career. Of necessity, the following account is a brief summary of Bull's work on this topic, and it attempts merely to capture the main themes that distinguished Bull's conception of international society from other conceptions, and helped to establish Bull's conception in the dominant position that it currently enjoys.

For Bull, '[t]he starting point of international relations is the existence of states.'[49] He then argued that the concept of international system (understood to mean a system of states) is broader than that of international society, but that international society, in his use of the concept, 'presupposes an international system.'[50] Therefore, international society operates within a conceptual terrain already depicted in terms of states and systems of states. This presupposition inevitably rules out other answers to the question of the nature of international society:

> the complex of rules that states what may be called the fundamental or constitutional normative principle of world politics in the present era... identifies the idea of a society of *states*, as opposed to such alternative ideas as that of a universal empire, a cosmopolitan community of individual human beings, or a Hobbesian state of nature or state of war, as the supreme normative principle of the political organisation of mankind.[51]

Thanks to his chosen starting point, Bull was then in a position to go on to give a specific account of the main normative principles embedded in international society in terms of reciprocity, sovereignty and non-intervention. This list excluded alternative elements that other conceptions might have included in the normative structure of international society, especially aspects of cosmopolitanism or liberal constitutionalist themes located by Wight and others in 'Western values'. In contrast, Bull argued that 'international society is quite inhospitable to notions of cosmopolitan justice, and able to give only a selective and ambiguous welcome to ideas of human justice.'[52] Thus, as Bull himself explicitly recognised, his limited ascription of membership undermines alternative conceptions of international society:

> the idea of international society identifies states as members of this society and units competent to carry out political tasks within it... it thus excludes conceptions which assign this political competence to groups other than the state, such as universal authorities above it or sectional groups within it.[53]

A good example of this can be found in Bull's discussion of two different conceptions of international society: a pluralist one, in which states 'are capable of agreeing only for certain minimum purposes'; and a solidarist or Grotian one, which assumes that states are capable of collective action to enforce the law and in which 'the members of international society are ultimately not states but individuals.'[54] Now, while these appear to offer a rough version of the conceptual contest described earlier, this disagreement is conducted on terms which largely accept that modern international society is institutionalised on the basis of states and state practices. Even the Grotian conception, on Bull's formulation, refers to 'the solidarity, or potential solidarity, of the states comprising international society.'[55]

Bull's defence of his definition of international society: Grotianism and modernity

Thus far, we have seen that Bull recognised, in a rather limited way, the existence of a dispute about the concept of international society, but that he definitely preferred one particular conception of international society over others. He backed up his preference with two arguments. The first was a re-working of the different traditions of international political and legal thought, in which Realist-positivism became 'Machiavellianism', Rationalist-eclecticism became 'Grotianism' (on Bull's own terms), and Revolutionist-naturalism became 'Kantianism'.[56] The second argument was an historical account of the particular values, rules and institutions that comprise modern international society. Let us now look at each of these arguments in turn.

In the first argument, Bull in effect practised two sleights of hand. First, he developed what purported to be an account of Wight's three traditions, but in which, unlike Wight's account, the concept of 'international society' was tied exclusively to the Grotian/Rationalist tradition. Then, he subtly changed the content of the Grotian/Rationalist conception of international society to emphasise its statist dimension. On the first point, Bull faithfully recorded the principal doctrines of Realism, Rationalism and Revolutionism; but, crucially, he stopped short of describing the different traditions' views as different conceptions of international society. Thus, Bull's description of Machiavellians attributed to them the view that 'there is no international society; what purports to be international society... is fictitious.'[57] He did not go on to discuss the Realist-positivist conception of 'international society' that Wight had identified. Similarly, the Kantians' assertion of an essential community of mankind was related by Bull to a concern with 'international morality'. He did not mention Wight's point that this was also a vision of the real essence of international society in terms of relations between individual human beings, although it seems odd to suppose that one can meaningfully appeal to a particular version of international morality without having a distinct conception of international society.[58] In the end, Bull acknowledged Wight's survey of the traditions' differences in terms of

> the distinctive doctrines that each of them put forward concerning war, diplomacy, power, national interest, the obligation of treaties, the obligation of an individual to bear arms, the conduct of foreign policy and the relations between civilised states and so-called barbarians.[59]

It is telling that their various conceptualisations of international society were not on this list.

Bull's decision not to attribute conceptions of international society to the Machiavellian and Kantian traditions led him to the conclusion that only Grotians conceive of international relations in terms of an international society. Indeed, both Hobbesians and Kantians were held to be radically antagonistic towards the idea of international society.[60] It therefore seems only to be the Grotian tradition that 'describes international politics in terms of... international society.'[61] Conveniently, this allowed Bull to be regarded as a standard-bearer for the theory of international society as a whole. His work thus came to be viewed by no less a commentator than John Vincent as 'a classical statement of the rationalist or Grotian position on world order. Against the realist deniers of international society, and the revolutionary destroyers of it, Bull argued that order in world politics was dependent on the survival of international society.'[62] This 'classical statement of the rationalist position' was, in Vincent's view, not jeopardised by the fact that Bull's work also contained 'an implicit defence of the states-system.'[63]

Furthermore, Bull depicted the Grotian tradition as attached to a surprisingly statist account of the membership of 'international society'. For Wight, as we saw earlier, the Grotian conception included states, individuals and non-state actors as members of 'international society'. In his early essays, Bull accepted this point, but by the time of writing *The Anarchical Society*, he came to insist that 'the Grotians accept the Hobbesian premise that sovereigns or states are the principal reality in international politics; the immediate members of international society are states rather than individual human beings.'[64] This is one aspect of the significance of Bull's dual use of the term 'Grotian' to refer both to the concept and to one particular conception of international society.[65] By relating one aspect of Grotianism to the idea of a society of states, the individualist, or at least non-statist strands of Grotian thought, which re-emerged partially in the idea that a 'solidarist' society of states would uphold individual rights, were muted, seen as prematurely idealist or as nostalgically medievalist, and brought in only as qualifications to an essentially statist perspective.

Let us now turn to Bull's second argument for his conception of international society: his historical account of the development of the particular values, rules and institutions that comprised the element of society in modern world politics. Here, once again, Bull acknowledged the existence of alternative conceptualisations of international society.

International Society: An Essay in Political Argument   33

These alternatives were, however, introduced only to be dismissed as irrelevant to the way in which modern international relations are organised. His claim in this context was that the conception of a society of states has a special practical significance in modern international relations. As he said in his defence of the representation of international society as a society of states, which was quoted above, 'the complex of rules that states what may be called the fundamental or constitutional normative principle of world politics *in the present era*... is the principle that identifies the idea of a society of states, as opposed to... alternative ideas.'[66] The crucial part of this claim is, of course, historical: Bull's conception was intended to apply to the 'present era'. His account therefore depended on an historical account of the nature of modern international relations.

Bull began by using the concept of international society to describe sixteenth and seventeenth century 'Christian international society', Europe in the eighteenth and nineteenth centuries, and contemporary 'world international society.'[67] However, his discussion of the 'Christian international society' was couched almost entirely in negative terms. This society lacked a clear membership principle, it had inchoate rules of coexistence, and it had no properly defined institutions.[68] It is, in fact, rather puzzling on what grounds Bull used the concept of international society to describe this arrangement at all, since he obviously thought that it lacked many of the elements which he had previously depicted as essential to the idea of international society. One can only suppose that he did not think that, for example, an ascription of membership only to states is a necessary requirement for an international society, and that he actually thought that the essence of international society lies, after all, simply in the existence of certain normative principles towards which international conduct is oriented.

The transition to the modern European states system was, according to Bull, accompanied by the clarification, consolidation and increased sophistication of the rules and institutions of international society. Most importantly, the membership principle was properly established. In the eighteenth and nineteenth centuries,

> the ambiguity of earlier thinkers as to what kinds of groups or entities are members of the society of states [*sic*] gives way to a clear statement of the principle that international society is a

society of states – even though this is sometimes accompanied by a qualification... From this recognition that all members of international society are a particular kind of political entity called "states"... there stem other basic features of the idea of "international society" in this period, [without which] it could not have been conceivable...[69]

In this way, Bull framed the distinction between different conceptions of international society, in terms of an historical transition from medieval or early modern Christendom (an at best inchoate international society) to the much clearer rules and institutions of the modern European society of states. The result of this argument was that, in conjunction with the former treatment of the Grotian tradition, the contested nature of the concept of international society was framed in such a way that only one conception could possibly have been seen as relevant to modern world politics.

Bull's critics within the English school

Bull's way of framing the concept of international society and his defence of one particular conception of international society did not go unchallenged in the immediate intellectual context within which he advanced his ideas: the 'English school' of international relations theory. For reasons of brevity and clarity, we can treat this school as equivalent to the British Committee for the Theory of International Politics, a body of which Bull was a member, as were Wight, Butterfield and Adam Watson, among others.

When one looks at the debates that went on in the British Committee, one can see several instances where Bull's depictions of the Rationalist tradition and the Grotian conception of international society were questioned. For example, Butterfield argued that the concepts of Christendom and ultimate individual responsibility were central to Grotian thought, rather than the assumption of solidarism *between states*. This might seem like a mere change in emphasis, but it was one with very considerable implications for the theory of political obligation believed to hold in international society.[70] Another British Committee member, the theologian and philosopher Donald Mackinnon, made a further contribution to this debate that engaged more systematically with the distinction between natural law and natural rights. This moved the discussion towards the radical elements of early modern European political theory, where '[a] natural right is a right that men claim to be theirs as men; they demand recognition of their right to subsistence or freedom to associate, not as citizens... but as *men*.'[71] This is, as Mackinnon argued, both individualist and naturalist, and implies a voluntarist conception of political obligation in international society. In particular, Mackinnon called for a liberal theory of political obligation in international relations which, 'must engage with the very difficult question of the individual's justification in challenging and indeed resisting decisions in the field of international relations, made by the executive.'[72] In other words, Mackinnon in effect argued for a radical version of Grotian natural rights theory as the basis for a conception of international society constructed from a natural rights perspective, and which would justify individual resistance against sovereign authority and individual personality in international society.

The room for this debate within the English school was created by Hugo Grotius' articulation of two different political projects. In the apt phrase of Richard Tuck, 'Grotius was both the first conservative rights theorist in Protestant Europe and also, in a sense, the first radical rights theorist.'[73] This dualism created the possibility of constructing the concept of international society from two very different evaluative perspectives. One was consistent with individual rights, republican theories of political community and a theory of obligation including a strong right to resistance. The other was consistent with the idea of a society of states organised according to the more authoritarian view of politics embodied by the principles of absolute external sovereignty and nonintervention. Within the English school, debates about the meaning of 'Western values' and Rationalism were initially constructed along these lines. However, Grotius's political theory is an early, ambivalent, weak and imperfect account of voluntarism, compared with, for example, John Locke's.[74] By being forced to discuss international social norms in the specific context of Grotian political thought, therefore, Butterfield and Mackinnon were immediately put at a slight, but by no means insurmountable, disadvantage.

The tendency towards a more authoritarian theory of political obligation in international society was, however, made decisive by a further aspect of the English school's research programme: its attachment to a statist conception of modern European history. Probably the major task which the British Committee set itself was the comparative study of states-systems through history.[75] This choice was informed by the work of the 'Gottingen school' or German 'Historical school', including Alexander van Heeren. In this way, the English school's historical account of the norms, rules and institutions of modern international society came to be informed by the importance attached by Heeren to the Peace of Westphalia as a key moment within an ongoing process of development. Heeren saw this moment as crucial to the emergence of modern European international society because he saw the Peace as 'settling the leading political maxims' according to which the 'subsequent policy of Europe' was conducted.[76] In particular, these maxims were embodied by the German constitution clarified in the Westphalian Treaties, which established the independence of the German states with respect to the Holy Roman Emperor and thus helped to make the balance of power a principal object of state conduct.[77] The result of this emphasis on the independence of the German states under the Westphalian Treaties was that the 'most important' feature of modern European history came to be seen as the interaction of states in a system characterised by '*internal freedom*; that is, the stability and mutual independence of its members.'[78]

This played into Bull's hands in two ways. First, it drew attention to the emergence of states in continental Europe, and drew attention away from other aspects of early modern international relations that would have been more helpful to those interested in alternative conceptions of international society. The work of Charles Alexandrowicz on the affinities between the naturalist conception of the law of nations and early modern relations between European and Asian political communities is a fine example of how important such an alternative focus might have been in the defence of an alternative conception of international society.[79] Secondly, Heeren's account of modern European history was profoundly hostile to, and dismissive of, republican ideas and institutions. This attitude towards republicanism was integral to his understanding of the European states-system, because he used the supposed predominance of monarchical institutions to argue that 'the management of public affairs became more and more concentrated in the hands of princes and their ministers, and thus led to that cabinet policy which particularly characterises the European states-system.'[80] The pejorative treatment of republicanism helped to establish a bias in favour of treating Grotian solidarism as a pre-modern residue of medieval Christendom.

POLITICAL ARGUMENT IN INTERNATIONAL RELATIONS THEORY

The argument of this chapter might be summarised as follows. The concept of international society refers to a core idea that international relations are ordered or rule-governed, in the sense that it supposes international behaviour to be oriented towards normative principles. This basic concept is open to a number of different reasonable interpretations, since the question of the content of the normative principles to which orderly international behaviour is oriented involves the appraisal of a varied and complex collection of different aspects of international relations. Therefore, it is not, *prima facie*, unreasonable for a student of world politics to maintain that international society consists in something quite different from Bull's conception of an 'anarchical society' of formally free and equal, non-intervening sovereign states engaged in diplomatic intercourse with one another. The contested character of the concept of international society is not widely recognised in contemporary international relations theory because Bull was able to give a defence of his particular conception that was overwhelmingly compelling within the context of the specific research programme devised by the members of the English school. Thanks to the clarity and forcefulness of his views, as well as the limited historiographical assumptions of the English school, Bull happened to win the argument at one particular place and time.

There is, however, nothing sacrosanct about either Bull's relatively authoritarian reading of Grotius or the Rankean predilections of the English school, and there is no reason to commit oneself to agreeing with Bull's conception to the exclusion of all others. As a conclusion to this argument, I want to analyse the consequences for international society theory of the success of Bull's conception, and I want to ask how the open and contested character of the concept of international society might be more fully realised.

As was indicated in the earlier discussion about the idea of essential contestedness, the way theorists use concepts is central to the way in which political arguments are conducted. Conceptual contests, if properly conceived and conducted, are part of an agonistic idiom for political debate, within which a tradition is allowed, or encouraged, to develop. This was the case with regard to speculation about the nature of international society, prior to Bull's resolution of the debates within the English school. One can find there a diverse and relatively broad tradition of thinking about international society, in which many different perspectives could come together in a constructive and mutually beneficial argument. In an agonistic political argument, one can defend one particular position without having to deny outright other positions. Indeed, the whole point of such an argument is that one ought to recognise the reasonableness of many different points of view with regard to the meaning of a concept. Wight's broad analysis of the development of international theory illustrates a classic feature of this reciprocal and tolerant mode of political argument: gradually there emerged, 'a confluence, a convergence, if not to say confusion of traditions.'[81]

Bull's framing of the concept and his defence of his conception proved to be so compelling that it had the effect of resolving the 'continuities of conflict' present in international society theory until then, on the terms of the conception of international society as an 'anarchical society' of sovereign states. Thus, theorists now see the concept of international society almost exclusively in terms of Bull's conception. Of course, this does not mean that all international relations theorists agree with Bull's assessment of the desirability of international order. The point is that alternative evaluative perspectives are now forced to express themselves through altogether different concepts of the way in which orderly behaviour in international relations is governed. A fine early (1978) illustration of this development can be seen in Vincent's account of 'moral order in world politics', which is built on an explicit contrast between two different concepts: international society (the society of states) and world society (the community of mankind).[82] This distinction framed Vincent's subsequent discussion of the issue of human rights. However, it is interesting to note that, instead of arguing that world society was emerging, he argued that the normative content of the goals pursued within the society of states were changing into something different from those proposed by Bull's conception of international order.[83] Here we see the old contested structure of the concept of international society struggling to break free from the iron frame imposed by Bull's account of the norms, rules and institutions of the society of states. What is lacking, however, is a proper tradition, and an agonistically conceived concept, within which such a contest could flourish.

A different example of a similar phenomenon can be seen in Andrew Linklater's attempt to explore the intersection between the Rationalist and Revolutionist traditions, and hence to 'radicalise' the Rationalist tradition.[84] Linklater has developed this project in part through an exploration of the theory of political obligation contained in the modern international society of states, and in part through an analysis of the new forms of 'post-Westphalian' citizenship that appear to be emergent in contemporary world politics.[85] It would be helpful to put these interesting points about the diverse international political forms of obligation and citizenship into the context of an agonistic tradition of thought which did not depict modern international society exclusively in terms of legal and diplomatic relations between sovereign states. If we could draw on a broader range of meanings of the concept of international society to describe more features of norm-governed behaviour in modern world politics, we could more easily explore the affinities between the different normative principles embedded in Linklater's and Bull's understandings of the nature of modern international society. We could also interrogate more deeply the way in which these different norm-governed aspects of international relations were institutionalised in modern world politics, perhaps in ways other than through sovereign statehood and state practices.

This may seem to imply that the consequences of Bull's framing of the concept of international society have been largely negative, but that is not the case. By undermining the vitality of the tradition of international society theory, Bull (perhaps inadvertently) opened up the terrain of speculation about the element of society in international relations to encompass political perspectives that found no room for themselves within the previously established 'three traditions'. Thus, discourse about the norm-governed character of international behaviour is no longer confined to Realists, Rationalists and Revolutionists. Overall, then, any assessment of the consequences of this development in the way that the concept of international society is used should recognise that a trade-off has happened. The field of enquiry has been expanded considerably, but at the cost of losing an agonistic, and hence relatively reciprocal and cooperative, terrain for debate between different perspectives. A great deal more can now be said about what constitutes orderly international relations; but there are many more instances where people are simply talking past one another, and it is harder to synthesise these different perspectives of the norms that govern orderly international conduct.

In my view, it would be desirable to re-establish the agonistic mode of political argument about the nature of international society, but within a tradition conceived on more inclusive terms than was previously the case. A crucial first step in this direction is to recognise that there actually was once a 'living tradition' of speculation about international society within which that concept was contested, and hopefully this chapter will have helped to raise awareness of this. The next important step would be to loosen Bull's defining grip on the concept of international society, which would involve going back to the areas where Bull effectively won the argument about the proper way of conceiving of international society: his reading of Grotianism and his historical account of the nature of modern world politics. There is an excellent opportunity to challenge Bull on these issues, since, as was noted earlier, Grotius was an extremely ambiguous thinker, and it goes without saying that the concept of modernity affords plenty of room for rethinking what we take to be the significant components of the normative and institutional dimensions of world politics 'in the present era'. Once we appreciate that the idea of an 'anarchical society' of sovereign states is not a definition but an interpretation of the concept of international society, it might be possible to think about the question of the nature of international society in a more agonistic way from a wider variety of perspectives. This putative tradition would, admittedly, never be all-inclusive, but it might offer a conceptual terrain upon which a theorist could posit the importance of, say, liberal principles of justice in international society, while still accepting the reasonableness of interpretations that draw attention to other norms to which orderly behaviour in world politics is oriented, such as emancipation, power/knowledge or patriarchy. This would, hopefully, help us to move towards a better appreciation of what it means to speak of norm-governed conduct in international relations in general. It would also underwrite a form of 'agonistic pluralism' in international relations theory, in which political arguments were conducted between adversaries rather than enemies, and through which the settlement of disputes might more readily co-exist with the constant unsettlement of received orthodoxies.

NOTES

1. An earlier version of this chapter was presented at the *Millennium* 25th Anniversary Conference, and I would especially like to thank João Marques d'Almeida, Alejandro Colás, Eivind Hovden, Justin Rosenberg, Nicholas Wheeler and Peter Wilson for their very helpful comments and criticisms. I would also like to acknowledge the financial support of the Economic and Social Research Council for my PhD research, of which this chapter is a part.
2. William Connolly, *The Terms of Political Discourse*, Third Edition (Oxford: Blackwell, 1993) p. 227.
3. Chantal Mouffe, *The Return of the Political* (London: Verso, 1993) p. 4.
4. See Hedley Bull, *The Anarchical Society: A Study of Order in World Politics* (London: Macmillan, 1977); R.J. Vincent, *Nonintervention and International Order* (Princeton, NJ: Princeton University Press, 1974); Adam Watson, *The Evolution of International Society* (London and New York, NY: Routledge, 1992); and Hedley Bull and Adam Watson (eds.), *The Expansion of International Society* (Oxford: Clarendon Press, 1984).
5. Timothy Dunne, 'International Society: Theoretical Promises Fulfilled?', *Cooperation and Conflict* (Vol. 30, No. 2, 1995) pp. 145–6.
6. See, for example, Richard Falk, 'The Interplay of Westphalia and Charter Conceptions of International Legal Order', in Falk, Friedrich Kratochwil and Saul H. Mendlovitz (eds.), *International Law: A Contemporary Perspective* (London: Westview Press, 1985) pp. 116–42.
7. See, for example, Martin Shaw, *Global Society and International Relations* (Cambridge: Polity Press, 1994), and Ronnie D. Lipschutz , 'Reconstructing World Politics', *Millennium: Journal of International Studies* (Vol. 21, No. 3, 1992) pp. 389–420.
8. Ole Waever, 'International Society – Theoretical Promises Unfulfilled?', *Cooperation and Conflict* (Vol. 27, No. 1, 1992) p. 121.
9. R.B.J. Walker, 'Social Movements/World Politics', *Millennium: Journal of International Studies* (Vol. 23, No. 3, 1994) p. 694.
10. *Ibid.*, p. 695.
11. See W.B. Gallie, *Philosophy and the Historical Understanding* (London: Chatto & Windus, 1964) especially Chapter 8.
12. *Ibid.*, p. 156.
13. *Ibid.*, p. 161.
14. Connolly, *op. cit.*, p. 29.
15. *Ibid.*
16. Gallie, *op. cit.*, p. 161.
17. Andrew Mason, 'On Explaining Political Disagreement: The Notion of an Essentially Contested Concept', *Inquiry* (Vol. 33, No. 1, 1990) p. 83, emphasis original.
18. Gallie, *op. cit.*, pp. 161 and 168.

19. Steven Lukes, 'Relativism: Cognitive and Moral', *Transactions of the Aristotelian Society* (Supp. Vol. 48, 1974) p. 187.
20. See Felix Oppenheim, *Political Concepts: A Reconstruction* (Oxford: Basil Blackwell, 1981).
21. H.L.A. Hart, *Essays in Jurisprudence and Philosophy* (Oxford: Clarendon Press, 1983) p. 25.
22. Alasdair MacIntyre, *After Virtue: A Study in Moral Theory*, Second Edition (London: Duckworth, 1985) p. 222.
23. *Ibid.*
24. David Miller, 'Linguistic Philosophy and Political Theory', in Miller and Larry Siedentop (eds.), *The Nature of Political Theory* (Oxford: Clarendon Press, 1983) p.51. For a similar analysis, see also Christine Swanton, 'On the "Essential Contestedness" of Political Concepts', *Ethics* (Vol. 95, No. 4, 1985) pp. 811–27. For a reply to this objection see Connolly, *op. cit.*, pp. 226–27.
25. Martin Wight, *International Theory: The Three Traditions* (Leicester: Leicester University Press, 1992) p. 30.
26. *Ibid.*, p. 36, emphasis omitted.
27. *Ibid.*, pp. 36–7. See also, Martin Wight, 'Western Values in International Relations', in Herbert Butterfield and Wight (eds.), *Diplomatic Investigations: Essays in the Theory of International Politics* (London: George Allen and Unwin, 1966) pp. 101–2.
28. Wight, *International Theory, op. cit.*, p. 45.
29. *Ibid.*, p. 36.
30. *Ibid.*, p. 39.
31. Wight, 'Western Values in International Relations', *op. cit.*, p. 93.
32. Wight, *International Theory, op. cit.*, p. 48.
33. I am indebted for this point to Nicholas Wheeler.
34. See, for example, Amos Hershey, *The Essentials of Public International Law* (New York: Macmillan, 1912) p. 59.
35. John Westlake, *Chapters on the Principles of International Law* (Cambridge: Cambridge University Press, 1894) p. 3.
36. See, for example, Hersch Lauterpacht, *International Law: Volume I, The General Works*, ed. E. Lauterpacht (Cambridge: Cambridge University Press, 1970) p. 28.
37. Showing how Wight's account of the Realist, Rationalist and Revolutionist conceptions of international society is related to these different branches of international law anticipates a further possible objection to my claim that international society is an essentially contested concept. As we saw earlier, W.B. Gallie's account of conceptual contests located such disputes *within* a tradition; Wight's account of a conceptual dispute *between* traditions might therefore seem not to be evidence of a proper conceptual contest (I would like to thank João Marques d'Almeida for making this point clear to

me). However, if it is true that Wight's three conceptions of international society are contained within a broad tradition of speculation and debate about the nature of international law, I submit that this objection loses a great deal of its force.

38. Gallie, *op. cit.*, p. 170.
39. Chris Brown, 'International Theory and International Society: The Viability of the Middle Way?', *Review of International Studies* (Vol. 21, No. 2, 1995) p. 186, emphasis omitted.
40. *Ibid.*, p. 186n.
41. James Mayall, 'International Society and International Theory', in Michael Donelan (ed.), *The Reason of States* (London: George Allen & Unwin, 1978) p. 122.
42. Alan James, 'System or Society?', *Review of International Studies* (Vol. 19, No. 3, 1993) pp. 280–81.
43. R.J. Vincent, 'Order in International Politics', in J.D.B. Miller and Vincent (eds.), *Order and Violence: Hedley Bull and International Relations* (Oxford: Clarendon Press, 1990) p. 58, emphasis omitted.
44. Bull, *The Anarchical Society*, *op. cit.*, pp. 16–19.
45. These concerns recur frequently in Herbert Butterfield's work on international relations. One good illustration is Herbert Butterfield, 'The Balance of Power', in Butterfield and Martin Wight (eds.), *op. cit.*, especially pp. 141–42. However, the idea crops up in more developed forms in many other of his papers for the *British Committee on the Theory of International Politics*, such as 'Misgivings about the Western Attitude to World Affairs' (1959), 'Comments on Hedley Bull's Paper on the Grotian Conception of International Relations' (1962), 'The Historic "States-Systems"' (1965), and 'Toleration in Early Modern Times' (n.d.).
46. Wight, 'Western Values in International Relations', *op. cit.*, p. 89.
47. *Ibid.*, pp. 113 and 116.
48. On the idea of an 'English school', see Roy Jones, 'The "English School of International Relations": A Case for Closure', *Review of International Studies* (Vol. 7, No. 1, 1981) pp. 1–13; Sheila Grader, 'The English School of International Relations: Evidence and Evaluation', *Review of International Studies* (Vol. 14, No. 1, 1988) pp. 29–44; and Peter Wilson, 'The English School of International Relations: A Reply to Sheila Grader', *Review of International Studies* (Vol. 15, No. 1, 1989) pp. 49–58.
49. Bull, *The Anarchical Society*, *op. cit.*, p. 8, emphasis omitted.
50. *Ibid.*, p. 13.
51. *Ibid.*, pp. 67–68, emphasis added.
52. *Ibid.*, p. 90.
53. *Ibid.*, p. 68.
54. Hedley Bull, 'The Grotian Conception of International Society', in Butterfield and Wight (eds.), *Diplomatic Investigations*, *op. cit.*, p. 52.

55. *Ibid.*
56. This alternative scheme is outlined in Hedley Bull, 'Martin Wight and the Theory of International Relations', the Second Martin Wight Memorial Lecture, *British Journal of International Studies* (Vol. 2, No. 2, 1976) pp. 101–16. It is reprinted as the Introduction to Wight, *International Theory, op. cit.*, pp. ix–xxiii. References here are to the latter.
57. *Ibid.*, pp. xi–xii.
58. *Ibid.*, p. xii.
59. *Ibid.*, pp. xii–xiii.
60. Hedley Bull, 'Society and Anarchy in International Relations', in Butterfield and Wight (eds.), *Diplomatic Investigations, op. cit.*, pp. 38–39.
61. Bull, *The Anarchical Society, op. cit.*, p. 26.
62. Vincent, 'Order in International Politics', *op. cit.*, p. 58.
63. *Ibid.*
64. Bull, *The Anarchical Society, op. cit.*, p. 26.
65. *Ibid.*, p. 322, n. 3. The first of these two uses is in *ibid.*, p. 26. The second is developed in Bull, 'The Grotian Conception of International Society', *op. cit.*
66. Bull, *The Anarchical Society, op. cit.*, p. 67, emphasis added.
67. *Ibid.*, pp. 27–40.
68. *Ibid.*, pp. 28–32.
69. *Ibid.*, p. 34.
70. Butterfield, 'Comments on Hedley Bull's Paper on the Grotian Conception of International Society', *op. cit.*, p. 12.
71. Donald Mackinnon, 'Natural Law', in Butterfield and Wight (eds.), *Diplomatic Investigations, op. cit.*, p. 80, emphasis original.
72. Donald Mackinnon, 'Political Obligations', *British Committee Paper*, 1964, p. 7.
73. Richard Tuck, *Natural Rights Theories: Their Origin and Development* (Cambridge: Cambridge University Press, 1979) p. 71.
74. See the comparative analysis of Grotius, Locke and Pufendorf in Karl Olivecrona, 'Appropriation in the State of Nature: Locke on the Origin of Property', *History of Ideas* (Vol. 35, No. 3, 1974) pp. 211–30, or 'Locke's Theory of Appropriation', *Philosophical Quarterly* (Vol. 24, No. 96, 1974) pp. 220–34.
75. See Martin Wight, *Systems of States* (Leicester: Leicester University Press, 1977) p. 22.
76. Alexander van Heeren, *A Manual of the History of the Political System of Europe and its Colonies, from is Formation at the Close of the Fifteenth Century to its Re-Establishment upon the fall of Napoleon* (London: Henry G. Bohn, 1873) p. 103.
77. *Ibid.* pp. 101–03.
78. *Ibid.*, pp. vii and 5, emphasis original.

79. See Charles Alexandrowicz, *An Introduction to the History of the Law of Nations in the East Indies: Sixteenth, Seventeenth and Eighteenth Centuries* (Oxford: Clarendon Press, 1967), and see also Wight, *Systems of States, op. cit.*, pp. 117–28.
80. Heeren, *op. cit.*, p. 7.
81. Wight, *International Theory, op. cit.*, p. 265. On the proximity of the three traditions' conceptions of 'international society', see also *ibid.*, p. 47.
82. R.J. Vincent, 'Western Conceptions of a Universal Moral Order', *British Journal of International Studies* (Vol. 4, No. 1, 1978) pp. 20–46.
83. R.J. Vincent, *Human Rights and International Relations* (Cambridge: Cambridge University Press, 1986).
84. Andrew Linklater, 'Rationalism', in Scott Burchill and Linklater *et al.*, *Theories of International Relations* (London: Macmillan, 1996) p. 111.
85. See Andrew Linklater, *Men and Citizens in the Theory of International Relations*, Second Edition (London: Macmillan, 1990), and 'Citizenship and Sovereignty in the Post-Westphalian State', *European Journal of International Relations* (Vol. 2, No. 1, 1996) pp. 77–103.

3. Culture and International Relations: A New Reductionism?
Fred Halliday

INTRODUCTION: CULTURE IN CONTEMPORARY POLITICS

The question of culture is one conventionally separated from the study of politics in general, and has an apparently contradictory place in international relations (IR). If one looks at standard textbooks on international relations the subject, from any theoretical starting point, culture appears to be almost wholly absent. Yet one does not have to go far in contemporary discussions of international relations to find mention of something called 'culture'.[1]

The importance attached to culture is based on the claim that it has political import, as a source of conflict, or a basis for political community, or as the very constitutive level of political life itself. In other words, the study of contemporary international relations involves discussion of culture, if not as that factor which constitutes such relations, then, at the minimum, as a necessary component of them. Culture may not be the determinant of, or an autonomous element in politics, but it is an essential part of it – be it the case of dominant hegemonic forces or of those challenging power with alternative projects. The starting point of this discussion is, therefore, an examination of how this argument for the link between culture and international relations is made. This is done first by reference to debates in the contemporary world, and then by consideration of a number of theoretical standpoints. It then proceeds to identify some of the issues raised by such a widespread invocation of culture. Finally, it suggests a particular clarification of the problem, involving the development of the thinking of Antonio Gramsci and the suggestion of a broader agenda for the study of culture and international relations.

CULTURE, CULTURE EVERYWHERE

At least five different issues in debate about the contemporary world have brought the question of culture and politics to the fore. First of all, culture is a constituent part of the debate on globalisation. Accounts of globalisation identify two rival processes – the creation of a set of global linkages and communities on the one hand, and the emergence of local, fragmentary, regional entities on the other. The argument is, therefore, that the process of globalisation involves not only the creation of a unified world market, and of an increasingly unified network of political institutions, but also of a transnational cultural space, in which instruments of globalisation (satellite television, the Internet, pop culture) erode pre-existing differences defined in national terms.[2] At the same time, within hitherto more homogeneous states, and across frontiers, a variety of new social and cultural entities is emerging, challenging existing categories of nation, tradition, history and identity.

In mainstream IR discussion, globalisation, therefore, both homogenises *and* fragments culture. However, this contemporary globalising diffusion of culture has, in the non-western world above all, provoked a very different approach to culture, one of rejection. This is embodied in the thesis of 'cultural imperialism'. It is evident in the way in which such bodies as CNN or more generically, 'western media', are denounced. In contrast to globalisation, which leaves this issue open, the cultural imperialism argument opposes current trends. At its extreme it involves a conspiracy theory in two parts: (a) that somehow the media of the USA and possibly other rich states are controlling the world and (b) that a major or even the main form of control in the world today is through media and western-dominated consumer culture. At its worst, an extreme form of nationalist conspiracy theory, in more benign form a sort of generalised neo-Marcusian argument on 'manipulation', this emphasis on the role of culture in oppressing weaker societies is itself global. This hostility to globalised culture can, therefore, be found in many Third World countries, but is also evident, from both left and right, within European states. It is even identifiable in paranoid sectors of the USA itself.

More broadly, culture is given prominence by the renewed emphasis placed in international relations on nationalism. Identity, patriotism, tradition, roots are all keywords of the 1990s, be this in peaceful areas, where there is a stress on cultural diversity and richness, linked, at times, to questions of economic and political sovereignty, or in areas of the world, such as the Balkans, where inter-ethnic conflict has exploded. There is plenty in the discussion of nationalism that has implications for international relations, both at the level of peace and security, and in the normative sphere, where it is argued that nationalism is a legitimate, indeed desirable, means of realising principles of democracy in the contemporary world.[3]

Beyond the argument that nationalism is important, there is the claim that it is in some sense fundamental, constitutive of the contemporary world. This approach sees the international system as above all one in which ethnic and inter-cultural conflict dominate: variants of this thesis can be found – the 'new middle ages', or a new inter-ethnic pandemonium, being cases in point.[4] While much of this perspective rests on the idea of a breakdown of state power, leading to a lawless, culturally chaotic state of nature, there is an alternative statist reading, the argument that with the end of the other great divisions of international relations – colonialism, the Cold War – we are now entering an epoch of globalised conflict based on culture. This is, most obviously, the argument advanced by Samuel Huntington on a 'clash of civilisations'. Huntington's argument can be summarised in two ways: a *positive* argument, that states or blocs of states will increasingly define themselves in cultural terms and conflict on that basis; a *negative* argument that, since conflict in the international system is inevitable, and since no other basis for conflict now exists, culture will fill the gap – as he cogently puts it, 'If not culture, then what?'[5] The most obvious instance of this is the supposed conflict between Muslim countries, misleadingly aggregated and rendered as 'Islam', and the west, but other variants, for example, Confucian-western, or, within Europe, Slavic-Orthodox versus Western Christian, can also be found.[6] Perhaps more than any other, Huntington's apocalypse has appeared to provide a new view of international relations: it is a perspective avidly welcomed by cultural nationalists the world over, people who would, otherwise, not be seen dead espousing the views of an American East Coast professor.

If Huntington's thesis highlights culture by its apparently prospective approach, making a projection of the future, the case for culture has, however, acquired validation from apparently very different sources, namely recent history, in this case the end of the Cold War. The question of how and why the Cold War ended allows, like any such major event, multiple interpretations and all the candidate paradigms are ready with their answers: realists with an emphasis on the arms race, political theorists with their stress on democracy, nationalism experts with an account based on ethnic revolt, and economists examining a declining Soviet performance. It is, however, at least equally arguable that the long-run erosion and final collapse of the Soviet system, and the comparable if superficially less dramatic shift in China, is a result of cultural factors, linked to generational and social change. Communist states controlled access to publications and people from the western world, and jammed radios, not just to keep out information, in the narrow sense, but also to block images of an alternative way of life. In time, the old insulation of these societies broke down; new, more educated, social groups came to the fore; the culture of the west, mediated via radio, consumerism and increasing personal contact, all undermined the communist system. It was said of the late Willis Conover, the Voice of America jazz critic, that his programmes were more effective than a fleet of B-29 bombers. The Chinese state, railing against the 'sugar-coated bullets' of consumerism, and the 'smokeless war' waged by China's enemies, has said much the same thing, as does the Islamic Republic of Iran, now obsessed with *bombardiman-i tablighati*, 'propaganda bombardment'. Any account of the collapse of communism and the end of the Cold War will be multi-factor, but within such an account culture can and should play an important role. This explanation is all the more dramatic because it is being applied in the context of a strategic, highly militarised, inter-state conflict, that would, conventionally, be seen as the favourite stamping ground of hard-nosed realism.[7]

CULTURE AND SOCIAL THEORY

Whatever their implications in the world of theory these accounts of the relation of culture to politics have been given impetus by events in the real world or, rather, interpretations purporting to explain these events. The emphasis on culture and international relations has, however, also been promoted by quite other developments over recent years, in the realm of ideas. In each case culture and politics are asserted as linked, yet the conclusions vary greatly.

First of all, we can recognise the role of culture in liberal international relations theory – broadly speaking, the argument that conflict of the old military kind is declining, and other forms of interaction develop, which erode the power of states and classic security concerns, i.e. economics and culture, as factors lessening conflict and creating a more interdependent and transnational world. Talk of open, democratic, as against closed, aristocratic and hence warlike diplomacy, movements of town twinning and educational exchange, a general association of cosmopolitanism with pacificism – these have been stocks in trade of liberal international theory since the early nineteenth century at least. Its current manifestations are to be found in such writers as Robert Keohane and Joseph Nye, James Rosenau and other theorists of globalisation. The underlying theme of the more transnational version of this argument is that culture erodes barriers between states and nations and is good for peace and cooperation. An alternative, still liberal but more statist approach, is that of the later Joseph Nye, with his argument that the USA's lead in 'soft power' – technology, culture, lifestyle – will ensure its continued world leadership.[8] Secondly, there is the emergence within international relations theory of trends that explicitly locate culture as a constituent force in inter-state relations.[9] These writers see the construction of states as reflecting identity and cultural specificity; much is made of the concept of 'collective identity' and of how this is formed through the interaction of states and societies. The origins of this approach lie in the appropriation by writers on IR of the ethnomethodological or constructivist theories of sociologists, or, in a parallel vein, the application of the Weberian concept of *Verstehen*, i.e. 'understanding' that seeks to identify the meaning of actions to those who perform them, rather than of neutral analysis to IR.[10] There are differences of emphasis here – the work of Michael Hollis and Steve Smith, on 'explaining' *and* 'understanding' in IR is less subjective than the 'constructivist' and 'reflexionist' approaches pioneered by Alexander Wendt, and the 'Copenhagen School' of Ole Waever and others.[11] In this latter work in particular, the claim is not that culture has become more important recently, with the rise of nationalism or the end of the Cold War, but rather that we can rewrite the history of states and their interactions through a cultural lens.

This interest in meaning, identity, 'reflexion' receives confirmation from a third theoretical source, one more appropriate to normative theory. For parallel to the study of the impact of ethnicity and nationalism on international relations, and the policy implications thereof, there has been a growing concern within political and moral philosophy about the role of culture and tradition in defining communities, and the import of this for liberal thought. The implications of all of this normative thinking about the international are evident enough: by challenging assumptions of a universal morality, and the possibility of judgements across frontiers, this strand in moral philosophy has raised supposedly intractable issues for IR theory. This is all the more so as those who might appear to have established a foundation for a normative liberal IR theory appear reluctant to draw the kinds of implication that others have sought.[12] It would almost seem that the concepts of culture, tradition, and community have been used to set up a roadblock on the path to a stronger international normative theory.[13]

Two other theoretical approaches have, however, also drawn attention to culture and its role in international relations. One is structuralism, or theories of imperialism: here in both the Marxist (Immanuel Wallerstein) or non-Marxist (Johan Galtung) forms, culture is seen, along with military power, economics and politics, as one of the mechanisms through which the dominant 'centre' establishes its control over the subordinated 'periphery'. Culture serves here, not as in liberal internationalist theory to promote equality between states, but rather to consolidate the differences, the structures of inequality, between them.[14] The other theoretical school that emphasises culture is postmodernism. This is hardly surprising insofar as postmodernism began as a trend in architecture and in literary criticism. It proposes both a radical contestation of the primacy of any form of social activity or construction (hence the call for a polyvalent, polymorphous, nomadic, 'non-privileging' perspective) and an emphasis on the centrality of the symbolic, the discursive, the cultural in the constitution of life and of social relations. With 'non-privileging' and the iconic combined, postmodernism thus both stresses the importance of culture in the constitution of the international system, and itself translates what others regard as non-cultural activities – war, diplomacy, peace-keeping, economic integration – into cultural forms, 'texts' which can be read as cultural objects and where a multiplicity of interpretations and constructions is possible.[15] If the central Foucauldian insight into human activity, including international relations, is that the discursive and the definitional is a form of power, then the point of the writings of others, such as Derrida, is that all of life is a text. Potentially, everything is, therefore, culture.

If each of these theoretical approaches posits a link between culture and politics, the nature of this link is, however, markedly diverse. For liberal transnationalism, culture helps to break down barriers between states, and is an *agent* of a desirable globalisation. For the constructivists and their associates, politics *is* culture, in that states should be understood as cultural constructs. For the normative theorists, on the other hand, culture dictates a political stance, it *enjoins* a politics of non-interference and 'respect'. For Marxists and structuralists culture is an *instrument* of domination. For the postmodernists all of social activity, politics included, is cultural in *form*.

MEANINGS AND CLARIFICATIONS

Faced with this plethora of usages, it might seem that culture has swept all before it: states are cultural constructs, their interests and policies are determined by tradition and values, relations between them are mediated by culture and limited by particularism, the recent past and future are dominated by cultural differences. At this point, it is worth making a few initial clarifications, however. One can begin by asking of these arguments in favour of culture some basic questions of the kind one would ask of any other apparent new assertion in IR, 'interrogating' them as current idiom has it.

The first is definitional, the need to sort out the different ways in which the term is being used. As Raymond Williams has pointed out, culture is one of the two or three most complicated terms in the English language. He quotes Herder, to whom we owe the anthropological view of different 'cultures': 'nothing is more indeterminate than this word, and nothing more deceptive than its application to all nations and periods.'[16] Obviously 'culture' is being used in the IR literature in several different ways. There is nothing surprising, or wrong, about that in itself – the same can be said about many familiar IR terms, such as state, nation, power, structure or system – provided we are clear which usage we are using. At least four different meanings of culture seem to arise: culture as 'artistic culture', i.e. literature, art, music, painting; culture as contemporary media – satellite TV, fashion, pop music, lifestyle; culture as 'civilisation' – i.e. long-term systems of value and meaning, such as language, and religion; culture in the sociological sense, as a set of values defining community, as in identity, tradition and in legitimating or challenging systems of power.

All four uses of 'culture' are aspects of relations between states but different explanations lay greater or lesser emphasis on them: thus while opponents of 'cultural imperialism' are most concerned with the first and second, and Huntington with the third, the moral philosophers, as well as the sociologists, are concerned with the fourth. So too, in the main, are students of international relations: deployment of culture in the sociological sense does not preclude study of the narrower forms, but assesses them in their social context. When we enquire about the relation of culture to politics, and specifically to international relations, we are above all concerned to examine how culture interacts with, and serves to shape and promote, the state system and associated global structures.

Secondly, it is obvious that while all these approaches talk of 'culture', the approaches themselves and the uses they make of 'culture', are very different. Liberal internationalism, such as the work of Keohane and Nye on interdependence, sees the rise of culture as an index of cooperation; this contrasts with the Marxist structuralist use of culture, which stresses conflict. Both contrast with the realistic structuralism of Huntington[17]; constructivism's break with realism is only partial, since Wendt for one is keen to stress that his starting point is Waltz's view of anarchy; primordial nationalisms and Huntingtonian 'civilisation' theories have underlying assumptions wholly opposed to post-modernism. Indeed while each approach invokes culture to prove its point it turns out that *every* paradigm can use culture for its own purpose, and indeed has a ready-made role for it to play. Equally, taking general epistemological premises as the starting point, rather than IR theoretical paradigms, we can see that postmodernists, constructivists and philosophical realists can all use culture to reinforce their argument. The least we can say is that 'culture' settles nothing: no paradigm can persuasively claim culture as its own – they all do.

This diversity of meanings, and their implications, is all the more evident if we apply to this discussion the questions which should be applied to every claim about novelty in IR: is it significant, in terms other than the tautologies of that particular approach, and is it really new? Or, more kindly, how new is it? The literature on culture, like too many alternatives, makes more of its evidence than is warranted: the point is not whether culture is present, visible, believed to be effective, but to what extent it operates independently of other forces.

Other illusions play a supporting role: the state is, as ever, declining in importance; great play is made of the 'social movements' that are challenging states and frontiers, but this theme, like comparable invocations of the 'grass roots' and of 'empowerment', is too often delusory. On the novelty issue, the same applies: of course there is much that is new, for example, the Internet or satellite TV, just as there is, say, in economic globalisation or the changing role of the state. But, as in these examples, history can provide some restraint on 'presentism'.

The least we can say is that there are many echoes of earlier thinking in the current discussions of culture. Two schools of thought that immediately come to mind are Arnold Toynbee and the *Annales* school. For Toynbee 'civilisation', in a rather grander sense than Huntington's use, constitutes an organising principle of history, the rise and fall of states being observed against the backdrop of the rise and fall of civilisations. For the *Annales* school, history is less than of the *événementiel*, the sequence of specific events, and more that of medium and long-term events, some economic and social, some ecological, but also cultural which shape events and states at a particular time. If historical events in the strict sense belong to the first category, and ecological and demographic shifts to the third, that of the *longue durée*, cultural processes are concentrated in the intermediate space of the *conjoncture* .[18]

This emphasis on culture in international history is matched by that in international political theory. This is evident from the most prominent discussion of the role of culture in British IR: the thesis that states are bound into a 'society' by a set of shared values and assumptions. This English School approach assumes a set of states, deemed to be constitutionally equal, who hold these values in common and seek to protect and promote them. A comparable approach to the culture of inter-state relations is provided by regime theory in both its neoliberal and cognitivist variants.[19] By contrast, in the nineteenth century the prevailing imperial view of international relations was one in which culture played a central part not to promote interdependence, but as part of an explicitly *unequal and warlike* system – the right, and duty, of the white races, superior to all others, to impose their ways on the rest of the world. It was a world-view shot through with cultural assumptions and implications for policy, in this case a belief in the importance of culture in the strength, or 'vitality', of nations and in the right to stamp your language, religion, cuisine, sport, architecture, etc., on the rest of the world. Huntington may or may not be right, but there is much in his argument that harks back to this ethos. In the eighteenth century, there was a widespread Enlightenment belief in the existence of a common set of values that different states might share. Again, different interpretations were placed on this shared observation: Kant thought that a growing awareness of common values could hasten the advent of universal peace, while Burke thought that the common 'manners' of dynastic Europe constituted a form of commonwealth now threatened by the French revolution. Rousseau, on the other hand, was concerned to stop all this and prevent a dangerous intermingling of cultures that might threaten the purity of his communities: the reason he idealised Corsica, a place he never saw fit actually to visit, was that as a small island it could have as little to do with the outside world as possible.[20] The first project for a united states of Europe was that of the Abbé St. Pierre, an appeal to the Christian kings to unite against the Muslim Ottoman foe. Much further back, in Aristotle, we find the most basic division of the international system of all – into those who shared a common, Greek, language, and the others, *barbaroi*, who did not.

If these echoes should induce an element of caution into how far we are discovering something novel in the contemporary IR discovery of culture, a parallel note of caution can be derived from the historical premises found in much of the literature on culture itself, Huntington being a good, relatively sophisticated, example. First of all, there is a tendency in much 'cultural' explanation to explain the present in terms not only of cultural entities but also of their long-run cultural continuity. Thus the British attitude to continental Europe, or Muslim attitudes to Jews, or European views of Muslims are explained in terms of some autonomous influence, or effectivity, of cultural forces and generalisations. This is, however, to beg the question, not only as to whether such forces or identities exist, but also, even if they do, as to why and how they should operate across centuries or even millennia. It is much too easy in talking about culture to jump onto the bandwagon of transhistorical continuity: such continuities may operate, but it is an open question whether they in fact do. Claims as to their effectivity have to show why and how this is the case. The sociology of reproduction is a necessary bridge for such claims to cross:

Culture or tradition is not something that exists outside of or independently of individual human beings living together in society. Cultural values do not descend from heaven to influence the course of history. To explain behaviour in terms of cultural values is to engage in circular reasoning. The assumption of inertia, that cultural and social continuity do not require explanation, obliterates the fact that both have to be recreated anew in each generation, often with great pain and suffering. To maintain and transmit a value system, human beings are punched, bullied, sent to jail, thrown into concentration camps, cajoled, bribed, made into heroes, encouraged to read newspapers, stood up against a wall and shot, and sometimes even taught sociology. To speak of cultural inertia is to overlook the concrete interests and privileges that are served by indoctrination, education, and the entire complicated process of transmitting culture from one generation to the next.[21]

Any claim as to the influence of the 'past' or of 'cultural tradition' on contemporary politics should pass the test of Barrington-Moore's stricture.

At the same time, those who advocate a role for culture in international relations tend to have another historical premise, to take culture as a given, an autonomous and clearly defined factor whose relevance they then advocate – the moral philosophers, as well as Hungtington, are, in different ways, examples of this. Their arguments begin with an unspecified, and contestable, historical premise, that somehow 'cultures' exist and can be identified, almost like features on a map. Nineteenth century attempts to combine mapping of languages with mapping of races reappear here in a new form. But the lesson of much of the literature on culture is, of course, very different: here we see an awareness that culture is (a) manipulated by interest groups for their own purposes and (b) defined and often invented for precisely this end.[22] In other words what appears as a given culture, tradition, identity, or history is, in fact, created by contemporary political and social actors for their own purposes.

The literature on nationalism is replete with examples of this selection, and creation, of culture, as is that on tradition.[23] At the same time much that presents itself as distinct or traditional is, on closer examination, a reproduction of modular international themes. If one looks at supposedly 'indigenous' contemporary cultural creations – Islamic fundamentalism, or 'neo-Confucian' statism in the Far East – the same applies: the organising principles are universal, part of contemporary world discourses on sovereignty, non-interference, equality of states and so on.[24] Culture is not an eternal, or ahistorical, force: the use of words like 'tradition' or 'primordial' conceals this – as they are intended to do, since the claim of every proponent of culture, as of nationalism, is precisely that this is somehow given and therefore cannot be contested. The question which this historicisation implies is therefore as much what use existing states and contenders for state power are making of culture, as that of how culture is an independent influence on politics, or how it is weakening the power of states.

Underlying these questions are, however, the theoretical and in some cases epistemological questions which each of the uses of culture brings in its train, and which, by dint of the specific focus on culture, may too easily be overlooked. If the supposition of transhistorical continuity is one underlying weakness, even weaker may be the transposition of ideas, developed for individuals, to social collectivities: this is what Newton long ago termed 'the fallacy of composition'. The constructivist approach is vitiated by this problem alone. Moreover, the use of the Weberian concept of *Verstehen*, for example, and theories based on symbolic interactionism or ethnomethodology, the progenitors of the IR constructivists of today, has been the subject of a long critical literature within sociology that, at the least, compels us to question its uncritical application to IR. The central point of such criticisms is that all theories which confine themselves to subjective meaning ignore the objective structures of power which shape that influence and determine their effectivity.[25]

Even more questionable is the reliance on the actual statements of social actors, that is on discourse as a text that may in some way provide a path to analysing action and meaning. For we know that, sometimes unwittingly but often wittingly, and not least in the world of international relations, statements are uttered which do not correspond to reality: twentieth century politics has given us plenty of examples of misleading discourse – *Arbeit macht frei*, 'People's Democracies', 'Separate but Equal'. To assess the veracity, but also the meaning, of such statements we have to look at other aspects of that actor, and at facts. Analysis of meaning and discourse are pertinent, *but only* when linked to a study of these broader contexts. Failure to do so leads to a world of unanchored subjectivity, a 'semiocentrism' that obscures more than it reveals.

If the discussion of culture poses such analytic questions, it also raises ethical issues of a kind that are also too often obscured, and deliberately so, by recourse to a supposed 'given'. For the ethical presupposition behind nationalism, religion and other invocations of tradition, and it is a presupposition our moral philosophers seem too happy to accept, is that we are faced not only with an identifiable cultural given, but also one that commands moral authority. There are many ways of saying this, some traditional, some more contemporary, but the justification of nationalism, and of communitarian ethics, involves such a claim.[26] Part of the challenge to this involves analytic, and empirical, issues: tradition, identity, 'roots', the 'authentic' are not actually 'given', nor indeed are they as discrete as their upholders would suggest. But this leaves open the question of what to do about ethical claims: where communities or states behave in such a way as to oppress others in the international system, or where those with power within communities – governments, the rich, men, teachers even – so define the tradition that other groups are oppressed and are either not permitted to articulate different definitions of the tradition, or must simply leave the community. If we are supposed to be silenced by the invocation of tradition, we need not be. The invocation of culture, in this context, need not occasion acceptance of the values defined as part of it, but also allows for the articulation of an alternative, non-communitarian, ethical position. To take this latter approach is to surrender not just on ethical principle, but also to a range of other questionable phenomena – parochialism, cultural nationalism and the like. Both the definition of culture and, even when defined, the authority given to culture therefore need to be questioned.

TOWARDS COMPARATIVE CONTINGENCY: A GRAMSCIAN APPROACH

The importance of culture in the study of international relations may, therefore, be compromised by a simplistic application of this concept and by neglect of theoretical issues latent in earlier discussions of culture. There exists, however, a large literature on culture that is based on different premises, and that seeks to relate culture in its various forms to concepts of power, structure, and interest. One line of enquiry to provide such an anchoring, and to meet some of the difficulties identified above, would be to address the discussion of culture and its implications for politics in the work of one theorist, the Italian Marxist Antonio Gramsci (1896–1937).[27] Gramsci's name has been invoked in two recent theoretical debates: one, directly political, on the strategy for socialist parties and movements in western Europe, the other, more theoretical, as part of the development of a critical theory within international relations and in particular with regard to international political economy.[28] Here, I shall try to take Gramsci's discussion of culture, as developed for Italian society, and explore its implications for international relations. The hope is to take further a research agenda, rather than provide a set of answers, but in so doing to offer one way through the confusions that now surround the debate.

Gramsci's view of culture and politics in western Europe can be summarised in the following manner. In the first place, a distinction can be made between those societies, classified by him as eastern Europe, where the rulers maintain position mainly by domination, i.e. coercion, and those in western Europe where it is maintained by hegemony, i.e. by a combination of coercion *and* consent.[29] Taking Machiavelli's image of the Centaur, a two-headed beast, he saw all societies as involving two forms of power.[30] In the western case consent is assured by what he terms 'hegemony' that is both a certain kind of social alliance, between the rulers and other social groups, but also a particular set of values that reinforce a given political and social system. Gramsci's usages change but one reading of his work is not so much that hegemony is counterpoised to coercion, or what he calls 'domination', but rather that, whereas in the eastern cases rule is by coercion alone or rather predominantly by coercion (since consent is found in even the most extreme dictatorships) in the Western cases, coercion remains at the centre of the system of rule but encased in a broader cultural system. Coercion is a last resort, or, in another idiom, the gold at the centre of a fiduciary monetary issue, while culture and consent reinforce and surround it. In other words, hegemony works by creating a cultural system that promotes consent and which serves to legitimate the position of the rulers.[31]

Gramsci distinguishes between two kinds of cultural hegemony, each pertinent to the end of social stabilisation: one is the diffusion of the values of the rulers, i.e. getting their culture accepted as natural or at least tolerable by all groups in society; the other is the creation and reproduction of non-hegemonic, subordinated or 'corporativist' cultures, values specific to oppressed groups which prevent them from articulating a challenge to the ruling groups. The latter is evident in colonial as well as social situations: order is preserved as much by consolidating the oppressed with their own folklore and folkways as by annihilating all cultural differences.

Gramsci examines two dimensions of hegemony and the means of opposing it. One is the appropriation and monopoly by the rulers of cultural tradition in order to deny access to that tradition to those with alternative social programmes. In Gramsci's view one of the tasks of opposing hegemony is to challenge prevailing definitions of the cultural and theoretical past, to appropriate it for other, non-hegemonic, purposes. He indeed believed that only when those opposed to the existing political system had successfully articulated their own hegemonic project, including in this an alternative reading of tradition, would it be possible to challenge established systems of rule. In line with this interest in classical and universal culture, Gramsci was also highly sceptical of challenges that in the end took a corporativist, i.e. self-defeating, form. A Sardinian who remained concerned about the inequality of north and south in Italian life, he also opposed what would today be seen as forms of Sardinian nationalism. He was particularly insistent that a provincial language like Sardinian, without a classical tradition, could not be used to articulate universalist, or counter-hegemonic, thoughts. That was the road to corporativist self-isolation. Any cultural alternative had to be at the level of the culture it was opposing, up to the challenge of power and the constitution of a new, emancipatory, hegemony. While opposing hegemonic cultures, he does not lapse into vapid adulation of every alternative or a feckless, but in the end corporativist, identity politics.[32] Equally, far from rejecting high culture as 'elitist', he worked to appropriate it for all social groups.

For Gramsci, therefore, culture is not separate from, let alone counterpoised to, systems of social and economic power, but part of the reproduction of such systems of power. His approach is consistent with neither theories of culture as autonomous or opposed to states, nor 'semiocentric' approaches of the post-modernist kind. For Gramsci the key to a successful hegemony is for the rulers to present their values, their view of history and what is appropriate, as natural, inevitable, and given. This theory is, therefore, consistent with those who analyse the instrumental, confected, nature of ideologies and of nationalisms and other cultural products in particular. Part of the challenge to such hegemony is, precisely, to question the naturalness of such values and the social implications they embody.

The implications of this for a critical study of the international system are many. In the first place, Gramsci's approach provides a means of seeing culture not as disembodied from power relations, let alone as an alternative to them, but as itself part of power, and of the conflict between social and political forces within and between societies. Culture, far from being a way of denying social reality, or the relevance of such categories as state, structure or class, is a means of better identifying their means of operation and reproduction. Similarly, what nationalist elites do is to create, through nationalism and the instrumental use of tradition and identity, particular forms of hegemony or, when they are out of power, counter-hegemony. This they do, above all, by seeking to present as natural, given, what is in fact of their own choice or creation. What is presented as an explosion of nationalism, or ethnic politics, or identity in the contemporary world can also be seen as a situation in which multiple rival groups competing for wealth and power use culture to mobilise and intimidate: this was very evidently the case with the emergence of post-communist nationalist elites in Yugoslavia or Transcaucasus. Huntington and others use the case of the wars in Bosnia and Croatia to illustrate their claims as to the role of culture, but other more instrumental explanations are available. More broadly, culture in the contemporary world and in international relations cannot be divorced from the interests of those with power – political and economic.

Secondly, it can be argued that the distinction Gramsci drew, between the eastern and western forms of rule, corresponds broadly to a narrow (pre-1945) and universal (post-1945) concept of international society. While colonialism certainly involved an element of cultural hegemony, diffusing elements of the metropolitan culture to subjected peoples, it rested predominantly on coercion, directed and mediated via collaborating local elites. The contemporary system of states rests on the appearance of equality, and naturalness, embodied in the United Nations and a world of sovereign states. Yet despite this appearance of equality, one to which all states pay tribute, a deep structure of inequality exists between states and between societies. This inequality is, however, masked by the appearance of equality and, through the workings of a hegemonic system, through the diffusion of the values of the powerful states to others.

Such a diffusion involves not just the acceptance of political values pertaining to 'international society', but also to the broader diffusion of the cultures of these states. Such a diffusion takes many forms, but in broad terms we see the creation of a hegemonically defined international civil society based on the power of western cultural norms. Beneath the international security system and a well-defended system of economic structures lie assumptions about the inevitability and naturalness of current social and economic orders – even claims, as in Francis Fukuyama, that history, in the sense of meaningful dispute about ideas, is over. Culture in every sense is part of this hegemonic order: the domination of the English language, promoted most recently through information technology, is a perfect example of an apparently natural, technical, process that embodies the supremacy, indeed near monopoly, of one society and its language over others. It is an open question, subject to empirical analysis, of what the combination of coercion and consent is in the contemporary international system, and what the role of culture, whether diffused by states or by the communications industry, is in this process. But that the promoters and process are mutually intertwined is indisputable. This is, in a limited sense, the thesis of the theory of 'soft power', articulated by Joseph Nye. It suggests questions about the international system that insert the study of culture into a broader study of hegemony and diffusion.

The implications of Gramsci's work for counter-hegemony are equally striking. One does not have to look far in the contemporary world to see cases both of hegemony in the sense of the acceptance of values by the subordinated, and of the adoption by the subordinated of values that, while rejecting hegemony, only confirm that system of rule. As already noted, the diffusion of one of the world's four thousand languages across the globe, and the acceptance as inevitable or natural of a certain definition of the market are examples of the first. Ideologies of submission, of fatalistic acceptance, or angry but self-defeating forms of revolt are examples of the second. So too are ideologies and movements that see in other subordinated peoples the main enemy – sectarian, communalist, tribal conflicts do little to alter the overall division of power. The world has indeed seen many such forms of revolt which, while appearing to be alternative forms of hegemony in fact lock their proponents even further into subordination. Cultural nationalism is an example of this, as are, in several countries, movements of religious fundamentalism: far from enabling their societies to escape the constraints of the international system they disable them, in part by rejecting precisely that reappropriation of classical and universal culture that is essential to a plausible hegemonic project. Equally corporativist are projects of economic autarchy, based on separating a country from the supposedly constraining world market, as are, in current UK debates, demagogic counterposings of a national economic policy to the entrapments of European integration. Those advocating such policies are shooting themselves in the foot: they do not challenge, but only fail to escape. The apparent tolerance of cultural revival, multiculturalism, identity politics, divorced from confrontation with material and other social structures of power, which is a hallmark of the contemporary world, may also promise a false emancipation, since they are in essence corporativist and limited in scope. It is precisely by confusing sign with reality, discourse with power structure, that this kind of alternative may reinforce the subordination that it is supposedly challenging.

CONCLUSION: THE AGENDA OF 'COMPARATIVE CONTINGENCY'

Such an approach can be termed that of 'comparative contingency': cultural values and ideologies are seen as contingent in two senses – as *dependent* on, or needing to be shaped, introduced and reproduced by, other social factors, and as *variant*, the variation being decided upon according to which rendering suits those with power. The comparison of different instances would enable broader judgements to be made, as well as to facilitate distinctions between histories and outcomes. One way to resolve these difficulties within IR would, therefore, be to take further the research agenda present in such work, one that was able to recognise the novelty of the contemporary context, but to build on work already done in many other parts of the social sciences.

Such an agenda might encompass the following:

1. The comparative study of culture in systems of power: i.e. the role of culture in establishing and reproducing different forms of power. This could encompass the role of, for example, education, language policy, religion, symbolism, national holidays in the constitution of modern states; the role of education and language in colonialism; the uses of culture by authoritarian states; the role of culture in contemporary hegemony, most specifically that of the USA.

2. Instrumentalism and culture: the ways in which states, or those aspiring to take hold of states, have used and indeed defined cultures to meet their political ends and articulate interests. This is the basis of, for example, some fine work on Islamism, both as anti-state and statist, ideology.[33] It also pertains to the promotion by states of particular interpretations of the past, and the legitimacy of political systems.

3. The comparative study of cultural transnationalism: the mechanisms of its diffusion and its impacts on states and society, prior to as well as including contemporary globalisation. Some recent work on globalisation has looked at culture as a domain for comparative historical analysis, but the study of how culture operates transnationally, now or in earlier epochs, remains elusive.

4. The role of culture in challenges to hegemony, and the successes and failures in this regard.[34] This was, as indicated, the central concern of Gramsci's work, the articulation of cultural alternatives as an essential part of any challenge to established regimes. Nationalism and fundamentalism have both sought to do this, as did the radical third world synthesis of Frantz Fanon.[35]

5. The comparative study of concrete cases of cultural collectivities, showing, without constructivist or transhistorical presuppositions, how such collectivities actually developed.[36] This would go for nations, social

communities, ethnic groups and transnational interest groups: each group, when it has emerged, may appear to be inevitable, a product of nature/God/destiny et. al. The challenge to the social scientist is to identify the elements of contingency, in the form such a group takes, and in the very fact that it does emerge at all, at both the domestic *and* international levels. As far as the latter is concerned, many nations owe their existence to the arbitrary occupation and border policies of colonialism.[37]

6. The development of the normative position that breaks with the constraints of nationalist, or communitarian, relativism. This has, in one sense, always been present, in the internationalist, and rational, critiques of nationalism. The recent revival of philosophical defences of nationalism, and similar anti-rationalist positions, has occasioned a reinvigoration of such ideas.[38]

None of these approaches would take it for granted that culture had an autonomous impact, nor would they assume that the relation between culture and the state or economy were constant as between different epochs. In the face of a tide of literature that is either analytically or epistemologically keen to establish its premises, these questions, and many others, should remain open.

Enough should by now have been said to indicate that in discussing culture in international relations we are not dealing with something that is, in a historical or theoretical sense, wholly new, nor are we dealing with a single object of analysis, let alone a single theoretical approach. Little wonder then that much confusion has surrounded discussion of the subject. We have discussions of the role of culture abstracted from the relations of power that produce and diffuse them. We have invocations of culture, tradition and the like which see them as historically given, without examination of who has created and promoted them and when. We have a defence of fabrication masking as history, proponents of petty-mindedness and moral agnosticism masquerading as defenders of moral probity. When it comes to historical overviews we encounter a free fire zone of half-truth and invalidated generalisation – on the one hand, theories of culture politics and media manipulation that are just globalised conspiracy theories, on the other, faultline babble that uses the past to present an ahistorical account of present and future. A degree of caution, and anchoring, would not now be amiss.

In talking of culture, we are dealing with a topic that is an essential part of any understanding of the international system, present from the beginning in one form or another in thinking on international relations, and now as much as ever promoted by a combination of practical and theoretical trends. Discussion of culture needs, however, to be embedded both in awareness of the history of ideas on this topic, and in the conflicting messages of history itself. Equally it presupposes an epistemology and a theory of power that, while recognising the importance of culture, avoids the illusions of semiocentrism and decontextualised overstatement. As Gramsci so cogently argued, culture is not an alternative to concepts of economic and political power, but a constituent part of their reproduction.

NOTES

1. There is, for example, no index entry for it in K.J. Holsti, *International Politics: A Framework for Analysis* (London: Prentice-Hall International, various editions).
2. Leslie Sklair, *Sociology of the Global System* (Hemel Hempstead: Harvester/Wheatsheaf, 1991); Michael Featherstone (ed.), *Global Culture: Nationalism, Globalization and Modernity* (London: Sage, 1990); Bertrand Badie, Marie-Claude Smouts, *Le retournement du monde. Sociologie de la scene internationale* (Paris: Presses de la Fondation Nationale des Sciences Politiques, 1992). Culture has been one of the core variables examined by David Held and Anthony McGrew in their major ESRC-funded study of globalisation (publications forthcoming).
3. For the implications of nationalism for international relations see James Mayall, *Nationalism and International Society* (Cambridge: Cambridge University Press, 1990); for a justification of nationalism on democratic grounds see David Miller, *On Nationality* (Oxford: Clarendon Press, 1995).
4. Alain Minc, *Le Nouveau Moyen Age* (Paris: Gallimard, 1993); Pierre Hassner, 'Nous entrons dans un nouveau Moyen Age' *Le Monde*, 27 October 1992; Daniel Moynihan, *Pandemonium, Ethnicity in International Politics* (Oxford: Oxford University Press, 1993).
5. Samuel Huntington, 'A Clash of Civilisations?', *Foreign Affairs* (Vol. 72, No. 3, Summer 1993), pp. 22–49 and debate in subsequent issues, later developed as *The Clash of Civilisations and the Remaking of World Order* (London: Simon and Schuster, 1997).
6. On the 'Islam' case I have tried to identify and disentangle some of the relevant confusions in *Islam and the Myth of Confrontation* (London: I.B. Tauris, 1996).
7. I have tried to go into greater detail on this in my *Rethinking International Relations* (London: Macmillan Press, 1994), chapter 5, 'International Society as Homogeneity' and chapter 9, 'A Singular Collapse: the Soviet Union and Inter-State Competition'.
8. Joseph Nye, *Bound to Lead* (New York, N.Y.: Basic Books, 1990), pp. 193–5.
9. Yosef Lapid and Friedrich Kratochwil (eds.), *The Return of Culture and Identity in IR Theory* (London: Lynne Rienner, 1996) gives an overview of this literature. In common with much of the culture literature, this discussion rather neglects recognition of the contributions of earlier thinkers.
10. For a sympathetic sociological introduction see Paul Rock, *The Making of Symbolic Interactionism* (London: Macmillan Press, 1979).
11. In addition to Lapid and Kratochwil (eds.), *op. cit.*, see Martin Hollis and Steve Smith, *Explaining and Understanding International Relations* (Oxford: Clarendon Press, 1991); Alexander Wendt, 'Anarchy is What States Make of

It: The Social Construction of Power Politics', *International Organization* (Vol. 46, 1991), pp. 395–421; Thomas Biersteker and Cynthia Weber, *State Sovereignty as Social Construct* (Cambridge: Cambridge University Press, 1996); Ole Waever (ed.), *Identity, Migration and the New Security Agenda in Europe* (London: Pinter, 1993). An earlier, most influential, general book on this approach was Peter Winch, *The Idea of a Social Science* (London: Routledge and Kegan Paul, 1958).

12. John Rawls, 'The Law of Peoples', in Steven Shute and Susan Hurley (eds.), *The Law of Peoples* (New York: Basic Books, 1993); Michael Walzer, *Thick and Thin: Moral Argument at Home and Abroad* (London: Notre Dame Press, 1994). The problem with so many of the moral philosophers is that while insisting that we are bound by our traditions, communities etc., they seem to be uncertain of what a community is.

13. The *reductio ad absurdum* of this trend is to be found in the work of Amitai Etzioni: a champion of 'community' at home, he denies the right to self-determination abroad. See Amitai Etzioni, 'The Evils of Self-Determination' *Foreign Policy* (Winter 1992-93), pp. 21-35.

14. Johan Galtung, 'A Structural Theory of Imperialism' *Journal of Peace Research* (Vol. 13, No. 2, 1971), pp. 81–119.

15. For the argument on the importance of culture in the formation of the international system see Edward Said, *Culture and Imperialism* (London: Routledge, 1992); for the latter see, *inter alia*, James Der Derian, *On Diplomacy* (Oxford: Blackwell, 1987). I have developed a critique of Said in *Islam and the Myth of Confrontation* (London: I.B.Tauris, 1996) chapter 7, 'Orientalism and Its Critics'. The unmediated application of techniques of literary criticism to social reality is itself questionable. These problems are also evident in the work of one of the most cogent and perceptive of postmodernist writers on IR, David Campbell: seeking to justify and illustrate the thesis that 'there is nothing outside of discourse' he analyses the Gulf War as performative, a virtual reality in *Politics Without Principle: Sovereignty, Ethics and the Narrative of the Gulf War* (London: Lynne Rienner, 1993).

16. Raymond Williams, *Keywords* (London: Fontana, 1976), p. 79.

17. Another realist usage of culture is the discussion of 'Power over Opinion' in E.H. Carr, *The Twenty Years' Crisis* (London: Macmillan Press, 1983).

18. Fernand Braudel, *A History of Civilisations* (London: Allen Lane, 1994) Part 1, 'A History of Civilizations'; Stuart Clark, 'The *Annales* School' in Quentin Skinner (ed.) *The Return of Grand Theory in the Human Sciences* (Cambridge: Cambridge University Press, 1985), pp. 177–198.

19. Andreas Hasenclever et. al., *Theories of International Regimes* (Cambridge: Cambridge University Press, 1997).

20. Stanley Hoffmann and David Fidler (eds.), *Rousseau on International Relations* (Oxford: Clarendon Press, 1991) part 6, 'Constitutional Project for Corsica'.

21. J.B. Barrington-Moore, *The Social Origins of Dictatorship and Democracy* (London: Allen Lane, 1967), p. 486.

22. For a cogent critique see Bill McSweeney, 'Identity and Security: Buzan and the Copenhagen school', *Review of International Studies* (Vol. 22, No. l, January 1996), pp. 81–93.

23. Ernest Gellner, *Nations and Nationalism* (Oxford: Blackwell, 1983); Benedict Anderson, *Imagined Communities* (London: Verso, 1982); Eric Hobsbawm and Terence Ranger (eds.), *The Invention of Tradition* (Cambridge: Cambridge University Press, 1983).

24. The work of Michael Leifer on 'Asian' values has most effectively illustrated this point.

25. Contemporary symbolic interactionism has tended to abandon the measured views of its founder, G.H. Mead. According to Mead, society has an objective existence and is not merely the subjective awareness of actors. It also fails to give sufficient weight to the objective constraints on social action. See the entry on 'Symbolic Interactionism' in Nicholas Abercrombie, Stephen Hill and Bryan Turner (eds.), *The Penguin Dictionary of Sociology* (London: Penguin, 1984). *The Penguin Dictionary* entry on 'Ethnomethodology' is equally apposite criticising this approach for, amongst other things, dealing with trivial subjects, and having no notion of social structure.

26. Miller, *op. cit.*

27. Gramsci's discussion of hegemony in Antonio Gramsci, *Selections from the Prison Notebooks* (London: Lawrence and Wishart, 1973), pp. 52–60. For elucidation in the secondary literature: John Cammett, *Antonio Gramsci and the Origins of Italian Communism* (Stanford: Stanford University Press, 1967); Giuseppe Fiori, *Antonio Gramsci* (London: Verso, 1969); Perry Anderson, 'The Antinomies of Antonio Gramsci', *New Left Review* (No. 100, November 1976 – January 1977), pp. 5–78.

28. Stephen Gill (ed.), *Gramsci, Historical Materialism and International Relations* (Cambridge: Cambridge University Press, 1993); Robert Cox with Timothy Sinclair, *Approaches to World Order* (Cambridge: Cambridge University Press, 1996), chapter 7, 'Gramsci, hegemony and international relations: an essay in method'.

29. Hegemony is 'an order in which a certain way of life and thought is dominant, in which one concept of reality is diffused throughout society in all its institutional and private manifestations, informing with its spirit all taste, morality, customs, religious and political principles, and all social relations, particularly in their intellectual and moral connotations,' Gwynn Williams, 'Gramsci's Concept of *Egemonia*', *Journal of the History of Ideas* (Vol. 21, No. 4, October–December 1960), p. 587, quoted in Cammett, *op. cit.*, p. 204.

30. 'You should understand, therefore, that there are two ways of fighting: by law or by force. The first way is natural to men, and the second to beasts. But as the first way often proves inadequate one must needs have recourse to the second. So a prince must understand how to make a nice use of the beast and the man'. Niccolo Machiavelli *The Prince* (London: Penguin, 1961), p. 99.

31. Gramsci is often associated with what has been termed the 'dominant ideology thesis' according to which the subordinate classes in society are deluded into accepting the values of their rulers. As the above hopes to show, this is not an accurate reading of his ideas, and more closely corresponds to the thinking of the Frankfurt School. See Nicholas Abercrombie, Stephen Hill and Bryan Turner, *The Dominant Ideology Thesis* (London: Allen & Unwin, 1980) and Goran Therborn, 'The New Questions of Subjectivity', *New Left Review* (No. 143, January–February 1984), pp. 97–107.

32. The concept of 'anti-systemic' movements, found in the work of Wallerstein and his associates, also underplays this distinction.

33. Ervand Abrahamian, *Khomeinism* (London: I.B. Tauris, 1993); Sami Zubaida, *Islam, the People and the State* (London: Routledge, 1989).

34. A good example of this kind of work is Graham Fuller, 'The Next Ideology', *Foreign Policy* (No. 98, Spring 1995), pp. 145–58, where he counterposes to Huntington the argument that movements of third world hostility are based on economic and social causes.

35. Frantz Fanon, *The Wretched of the Earth* (London: Macgibbon & Kee, 1965); and *Studies in a Dying Colonialism* (New York, N.Y.: Monthly Review Press, 1965).

36. This is, of course, the challenge facing all those who oppose perennialist or essentialist histories of nationalism. It *is* more difficult, more complex, but the others *are* wrong.

37. Two cases where I have tried to develop this with regard to nationalist movements: Eritrea, in Fred Halliday and Maxine Molyneux, *The Ethiopian Revolution* (London: Verso, 1981), pp. 175–82; Yemen, 'The Formation of Yemeni Nationalism' in Israel Gershoni and James Jankowski (eds.), *Rethinking Nationalism in the Middle East* (New York, N.Y.: Columbia University Press, 1997).

38. For a persuasive critique of relativism inherent in much post-modernist writing on gender see Sabina Lovibond, 'Feminism and Postmodernism', *New Left Review* (No. 178, November–December 1989), pp. 5–28.

4. Global Politics and the Problem of Culture: The Case of China
Christopher Hughes

The possibility that globalisation might result in a reduction of the power of the state has led a number of writers to ask what agencies might best be able to address international problems in a world of ever more rapid transactions.[1] As the possibility of a single world state remains either remote or undesirable, however, the various answers to this question tend to envision new forms of cooperation between states as the main way by which problems can be solved. This, however, immediately gives rise to what might best be called the 'liberal paradox', the problem that if states are to cooperate in ever closer ways, then they will have to share the same political culture. More specifically, because writers seeking solutions to global problems do not wish to do so at the expense of liberal-democracy, they argue that this will be a liberal-democratic culture. This, of course, immediately clashes with the cultural diversity that is evident in the world. It thus gives rise to an acute paradox for liberalism, as it sits uneasily with the conception of an international society premised on the toleration of diversity and the sanctity of sovereignty.[2]

If the liberal paradox is to be resolved, international relations needs to think more carefully about what kind of political activity can take place between different cultures. It will be suggested below that the starting point for this has to be a movement away from seeing liberal politics as the pursuit of fixed principles, or the realisation of a particular political ideology, in favour of seeing it as an *activity*. Probably the best account of this conception of politics is Bernard Crick's seminal defense of politics against challenges from ideologies of the right and the left.[3] This is an account of political activity that is rooted in the Hellenic tradition, drawing especially on Aristotle's *Politics*. What Crick finds most significant in the *Politics* is Aristotle's rejection of Plato's search for the ideal *polis* as a kind of unity. Instead, Aristotle observes that the *polis* is, and should be, an aggregate of its many members. If it becomes too unified it ceases to be a *polis*, 'as if you were to turn harmony into mere unison, or to reduce a theme to a single beat.'[4] The task of politics in such a pluralistic society should not be to eliminate rival ideological positions, but to create the conditions within which discussion can take place between them, an activity rooted in the Greek notion of dialectics.[5] A conception of political activity can thus be developed which 'represents at least some tolerance of differing truths, some recognition that government is possible, indeed best conducted, amid the open canvassing of rival interests.'[6]

When applied to the pluralistic international order, this conception of politics has obvious advantages. Before this is done, though, it must be pointed out that Aristotle argues for politics as a particular type of activity within the Greek *polis*, and Crick applies his arguments to the modern state, rejecting the possibility of political activity without a sovereign power. It will be argued below, however, that if we look carefully at how communication takes place across cultures, and with humanity facing common problems, increasing global transactions do lead to something akin to the kind of political activity that Crick describes. If this kind of discussion is to be more than just 'chickens talking to ducks', though, as a colloquial Chinese expression puts it, it has to be shown to lead to true communication – that is to some kind of shared meanings – rather than to either misunderstanding or mere acculturation. By drawing on developments in the social sciences, and especially sociology, it should be possible to indicate how this happens without getting stuck between cosmopolitan/communitarian and homogenisation/heterogenisation divides.

There is, of course, a danger that drawing on the wider social sciences may lead to a departure from the immediate political concerns of international relations. The discipline has, after all, suffered some withering attacks for straying too far from the real world.[7] One way to avoid this is to take a case study which raises some of the most acute problems for international relations and for globalisation. This is the case of China, a state that 'is remarkably integrated in the global economic system and steadily becoming more interdependent,'[8] yet which remains ruled by an authoritarian regime claiming to safeguard a 'socialist' system that is building a unique 'spiritual civilization'.[9] Whether we are talking about the global problems of pollution, migration, arms proliferation, or development, China's policies will have a decisive impact. The first part of the argument below will thus try to highlight some of the problems with recent analyses of globalisation by applying them to China. After that, some of the political implications for cross-cultural activity will be explored with reference to China's entry into international society. It is also to be hoped that such an approach will avoid another criticism laid at the door of some IR writers, that they can be historically short-sighted[10] and somewhat ethnocentric.[11]

CHINA AND THE LIBERAL PARADOX

The nature of the liberal paradox can be demonstrated quite clearly when the attempt by Paul Hirst and Grahame Thompson to rescue the state as political actor from the tyranny of global markets is tested against the case of China.[12] In brief, these authors see that it is desirable to maintain states as a kind of protective shield against the vicissitudes of the international economy, and argue that this is feasible because statistical evidence indicates that foreign direct investment (FDI), trade, investment and financial flows are concentrated in the G3 triad of Europe, Japan and North America, and even multinational corporations remain nationally based. This concentration of economic power in the G3 allows the authors to claim that there is a possibility for developing 'political strategy and action for national and international control of market economies in order to promote social goals.'[13]

Although such an argument is appealing to the G3, from outside this privileged group it can only be seen as a recipe for a new kind of hegemony. It is far from a truly global politics. This is already clear in the argument that the G3's economic power can be used for achieving political aims. It is even more so when these authors make normative judgements about what those aims should be:

> [d]emocracy, in the sense of representative government based upon universal suffrage, has become a virtually universal ideology and aspiration in the late twentieth century. Non-democratic regimes are now signs of political failure and chronic economic backwardness.[14]

Apart from the fact that such an assertion remains contestable on empirical grounds, little space is left by triumphalism for the plurality of political cultures that exist in the world. It is thus somewhat ironic that while these authors wish to defend the state against the tyranny of the markets, they end up asserting a vision of world politics in which market mechanisms are the main motor by which the G3 will transform other states. As Hirst and Thompson put it,

> [i]nward-looking nationalism and cultural fundamentalism are, to put it bluntly, the politics of losers. It is virtually impossible to operate in world markets and ignore at the same time the internationalised cultures that go along with them.[15]

In the sense of liberalism as the toleration of diversity, the implications of this vision for people in non-liberal democratic states are decidedly illiberal. How is such thinking any different from what American think tanks refer to approvingly as the 'tyranny of the markets' which is sufficiently 'corrosive of traditions and institutions in subtle ways' that it can be used to transform China?[16] Whether the prospect of universal liberal-democracy is to be lauded as the ideal form of global governance, or condemned for what Deng Xiaoping calls a 'smokeless war',[17] therefore, seems to depend very much on where one stands in the world.

The same kind of problem arises when global governance is envisioned through a kind of extension of domestic liberal-democratic theory to the international arena. Nowhere is this more evident than in the exploration of cosmopolitan democracy by David Held.[18] Aware of the diminishing ability of states to address transnational issues, Held argues for a world assembly in which inter-governmental organisations (IGOs), international non-governmental organisations (INGOs), citizen groups and social movements will be represented alongside states. Transnational problems will be resolved by referenda cutting across state borders, human rights issues will come under the compulsory jurisdiction of a new international court or near consensus voting in the General Assembly, and decisions would be enforced by a collective security force.

It is not difficult to see why such a global government would in principle be unacceptable to a state like China, which puts the highest premium on maintaining sovereignty over its domestic affairs.[19] In addition to the erosion of state sovereignty that this kind of system would entail, China, with most other states, would not even be eligible for membership of such an organisation, because, '[t]o begin with, at least, such an assembly is unlikely to be an assembly of all nations; for it would be an assembly of democratic nations, which would, in principle, draw in others over time.'[20] If these 'others' are to be drawn in though, they will have to accept the terms of a basic democratic law and be radically transformed, because 'a legitimate state must be a democratic state that upholds certain common values.'[21]

It is clear, then, that if global governance were to develop along the lines envisaged by the above authors, non-liberal-democratic states would either be faced with a tyranny of international markets used as a hegemonic tool by the G3, or with exclusion from an armed world government within which Amnesty International and Human Rights Watch Asia would be represented. For those concerned with the preservation of liberal-democracy, this is a laudable prescription, since, '[a]fter all, fascism, Nazism and Stalinism came close to obliterating democracy in the West only fifty years ago.'[22] Who, though, people outside the liberal-democracies might well ask, are the enemies of the West now? The Chinese certainly do not have to look any further than the works of Samuel Huntington or the cover of *The Economist* to have their worst fears confirmed.[23]

Arguments promoting liberal-democracy in a global context thus quickly bump up against the liberal paradox that arises from the toleration of diversity. Chinese rulers are fully aware of this inconsistency when they defend their record on human rights in terms of the specificity of their historical and cultural situation.[24] Held, it must be pointed out, is certainly aware of this paradox, realising that such a tactic is likely to gain increasing international support, as shown by events like the 1993 UN World Conference on Human Rights in Vienna.[25] More significant, though, is that this defense of totalitarianism and dictatorship even appears to hold credibility for many of those over whom Chinese leaders exert their rule, as is shown by the acceptance by even some prominent dissidents of the limits to political reform imposed for the sake of social stability.[26]

The more Held looks at the rise of the liberal-democratic state from the 'outside', then, the more he needs to qualify his vision of cosmopolitan democracy and accept that many states in the world reject political and civil-rights discourse along with Western dominance.[27] He shows some understanding of why this should be so when he points out that globalisation initially meant 'European globalisation', involving the erosion of the Muslim, Indian and Chinese civilisations, the destruction of the American Indians, the 'disorganizing effects of Western rule on a large number of small societies; and the interlinked degradation of the non-European and European worlds caused by the slave trade.'[28] Rather than enhancing liberty, the rise of the liberal-democratic state has thus involved 'great costs for the autonomy and independence of many, especially in smaller states and extra-European civilizations.'[29]

What is particularly interesting about Held, then, is that the more he moves away from his origins as a theorist of democracy and into the field of international relations, the more he comes to grasp the almost totemic value attached to sovereignty by post-colonial states, and the more he realises that the world has 'failed to break fundamentally with the logic of Westphalia.'[30] He thus feels that the 'transformationalist' view that globalising forces of interdependence are leading to a decline of the states system needs to be rejected, in favour of the conclusion that the 'nationalisation' of global politics has not run its course.[31] From the perspective of democratic theory, cosmopolitan democracy seems highly desirable; from the perspective of international relations it threatens the state sovereignty that is seen by many post-colonial cultures as the main guarantor of security from liberal-democratic hegemony. What is particularly interesting here, though, is that this problem forces all of the above writers on globalisation to turn to the cultural arena for a solution to the liberal paradox.

A CULTURAL SOLUTION TO THE PARADOX?

An example of this cultural turn can be found in Hirst and Thompson, who end up qualifying their hope for the emergence of an international civil society by approving of Kant's notion that the existence of different languages and religions will still guarantee that 'distinct local cultural traditions will continue to coexist with cosmopolitan cultural practices.'[32] Held can be seen moving in a similar direction when he points out that the future problem for theories of globalisation will be to grasp 'how and in what way cultures are linked and interrelated, through mutual accommodation, opposition, or resistance, for example, [and] not how a sealed cultural diversity can persist in the face of globalization.'[33] Unfortunately, though, despite this awareness of the need for a theory to explain the persistence of diversity in a global era, none of these authors actually attempt to develop their work any further in this direction. There is, however, a large body of literature on the nature of communication to which we can turn to fill this gap.

If we turn to hermeneutics, for example, the first observation that can be made is that understanding (and therefore political discussion) amounts to much more than a mere reproduction of one person's meaning in the mind of another. As has been argued from Dilthey to Gadamer, for example, the act of reading depends on 'making sense' of words within the context of rules and meanings that must be in the reader's mind before the whole text has been read.[34] The reader is thus conditioned by linguistic convention, but is also free to adapt that convention to create new meanings. Similarly, in the act of translation, the translator has to mediate between two linguistic worlds.[35] It would be pointless to reproduce the exact meaning in Chinese to an English audience, because this would not make sense without a deep knowledge of the text's original cultural and linguistic context. The only way to reproduce the original meaning would be to repeat the text in Chinese – a somewhat pointless exercise.

These observations about the creative nature of understanding are important to make for international politics because they indicate that cultures do not have to *either* become homogeneous *or* maintain incommensurable separate identities, for discussion to take place between them. Instead, they can be mutually involved in the generation of new meanings. When, for example, reformers at the end of the Qing dynasty (AD 1644–1911) had to explain the idea of a society of states, they had to do so by using a Chinese vocabulary with thousands of years of tradition attached to it. It is not surprising that new meanings, neither traditionally Chinese nor purely 'Western', should have emerged from this process.

Further insights into how novelty emerges from communication can be gained from the notion of 'speech acts'. From a Wittgensteinian perspective, for example, words are seen as having functions that are as varied as the tools in a tool-box.[36] Following on this, J.L. Austin points out that it is the intention, or 'illocutionary force' behind 'speech acts' that has to be grasped if their proper meaning is to be ascertained.[37] Habermas takes this point further and illustrates it when he draws attention to the variety of meanings that could be attached to somebody calling out the word 'attack'. Depending on how this word is being used, it might mean anything from an order to attack the enemy to a warning that an attack by the enemy is imminent.[38]

In other words, we can never know the meaning of a text from semantic analysis alone. Therefore, the likelihood through understanding as reproduction of original meaning decreases with historical and cultural distance. We need to be especially careful about this when we come across something familiar in another culture and assume that it bears the meaning that we commonly attach to it. The postmodernist writer Zhang Yiwu, for example, intersperses his Chinese text with English words. Knowing the meaning of these words in English is not so important for understanding why he does this, as knowing that his claimed intention is to make redundant the idea of a Chinese essential culture. His point is to create a new theory and vocabulary by using Western theory to criticise Chinese culture, and Chinese culture to test out Western theory. It would certainly be a mistake to see this as a mark of simple 'Westernisation'. This is, in fact, something for which he has been criticised in a Chinese book entitled *Cultural Criticism in a Pluralist Society*. The linguistic games become even more complex, however, when, with some aplomb, Zhang responds that every word in the title of his critic's book has actually transmigrated to China from the West via Japan.[39] Such an example alerts us to the danger of seeing familiar 'Western' concepts as evidence of Westernisation. Instead, 'Western' words can be used with the intention of both deconstructing Chinese culture and defending it.

To develop the political implications of this kind of inter-cultural language act somewhat further, we could turn to Habermas's theory of communicative action. This is because Habermas develops the theory of speech acts in a way that draws our attention to the fact that it is the power relationships within which those acts take place that determines their political meaning. In particular, he points out that if a person is to act through rationally motivated agreement, then she must do so from freely ascertaining the validity of a speaker's expression. Only if she can decide for herself if an utterance is true, or sincere, or legitimate, will the utterance have a 'binding effect'. The hearer will then be motivated to act without the need for coercion or bribery.[40] This means that for Habermas, '[t]he basic theoretical concept of the ethics of communication is "universal discourse"', which takes place in an 'ideal communication community'.[41] If we extend this to the international arena, what we have are the seeds of an explanation of the kind of environment that is needed if true global discussion is to take place. Ideally, it needs to take place in a context of equal power relations, and it cannot be expected to produce a homogeneous world culture.

GLOBAL PLURALITIES

Of course, when we are dealing with international relations we are looking at a communication community that is far from ideal in terms of power relationships. It can probably never achieve the ideal situation proposed by Habermas, but that ideal does have the heuristic benefit of constantly making us aware that power relationships do determine meaning to some degree. This point can again be illustrated well by the linguistic games involved in China's socialisation into 'international society' (*guoji shehui*). Here we have a concept that was originally developed from the intentions of sixteenth-century European jurists to make sense of relationships between states by drawing on existing conceptions of natural law as applied to individual human beings.[42] By the eighteenth century, lawyers had gone a step further. They began to juxtapose the idea of the state not just with the individual, but with the 'civilised' personality (i.e. the Christian state), by insisting that states could only become members if they measured up to a certain standard of civilisation.[43] In doing this, they had moved from a merely descriptive speech act to one with an illocutionary force intended to generate feelings of superiority and righteousness on the side of the civilised/coloniser, and inferiority on the side of the uncivilised/colonised. Analogy had become metaphor, the essential difference being that the latter has the specific function of stirring up emotions by juxtaposing familiar terms in unfamiliar ways.[44]

If this is so, then it is not surprising that when the concepts of international society are indigenised in China, we see that they get used in speech acts of resistance. The key nationalist text of the constitutional reformer Liang Qichao, *Xin Min Shuo* (*On the New People*),[45] produced in 1902, for example, opens by metaphorising the state as human body, with four limbs, internal organs, flesh and blood. Liang explains that it is created by a concentration of people, but if these people are stupid and timid, then the state will be like a body with broken limbs.[46] The state is thus not only explained, it is explained in such a way as to generate the emotions of indignation necessary to mobilise a population of 'villagers' who consider themselves to be living 'Under Heaven' (*tian xia*) rather than within a state.[47]

We thus begin to see how the transmigration of political ideas amounts to far more than merely reproducing or rejecting alien ideas. It is a matter of indigenisation through reinterpretation, a process that results in what has been called 'hybrid' concepts. Sociologists are aware of this when they look at how novel things emerge from combinations of the familiar with the unfamiliar, as with the example of Chinese-style tacos.[48] In global politics, hybrid concepts can play a key role in the development of new meanings. Take the case of how early Chinese nationalists like Liang Qichao, or his teacher, Kang Youwei, reinterpreted international society in terms of the Confucian canon.[49] Kang Youwei achieved this by locating international society as a stage on the path to the Confucian ideal of a stateless world of 'Great Harmony' (*da tong*), which provided a linguistic and conceptual bridge to the vocabulary of Marxist internationalism for Communists such as Mao Zedong.[50] Similarly, when Liang Qichao argued for a strong Chinese nation-state, it was to preserve thousands of years of the Chinese 'Way' (*dao de*) in a world dominated by the two great forces of conservatism and progressivism, by finding the right balance between them. What is notable is that, in formulating such an argument, Liang did not reject alien concepts, but had to interpret them in familiar terms that would make sense to his audience, such as the idea of the 'Anglo-Saxon race' as an entity that has managed to use 'one foot to stand, one foot to walk; one hand to hold and one hand to pick up.'[51]

Just as with the European jurists' approach, then, the Chinese state moves from being 'Under Heaven' to become an anthropomorphised applicant to international 'society'. The intentions and implications of this act are naturally very different in Europe and China. Alien concepts had to be indigenised in such a way as to mobilise a population thinking in a language with its own embedded conventions. The hybrid concept that results is encapsulated well within a manifesto issued by the Central Committee of the Nationalist Party (Kuomintang) in January 1934, two months before the Japanese annexed Manchuria, which explains the nature of the state as follows:

Now, there are two essential conditions of a modern State. In the first place, it must have a strong and compact organism; one in which there is good communication between its "head and tail" and coordination between its limbs and body. Secondly, it must be an organism which is strong and sound both within and without; that is, it must have healthy blood circulation within as well as well-developed muscles.

The lack of either, power without organisation or organisation without power, will render it impossible for such a state to maintain its independence externally or to protect its people internally. Thus, from the standpoint of political organisation, both unity and reconstruction are essential; while from that of the duty of the individual, it is necessary for the people to unite as well as to strive on.[52]

From this kind of language it is not hard to see that political discourse in China by the 1930s was not focussed on which party was best equipped to introduce liberal-democracy. The contest between the Nationalists and the Communists was over which party could 'save China' from Japanese imperialism by provoking and harnessing nationalist sentiments. For the eventual victors, maintaining their hegemony over the state-sponsored culture/ideology of Chinese nationalism remained a salient theme, as when ethnic minorities were encouraged to speak in their own languages so long as they talked about Marxism-Leninism.[53] The organic metaphor also remained evident, especially when the nationalist legitimation of the state was threatened by disunity, as in the case of Taiwan, when appeals for national unification had to be made on the grounds that the Chinese all share common 'bone and flesh'.[54] Such metaphors have become structural in the Chinese discourse of dissent. Deng Xiaoping defends China's 'state rights' as opposed to 'human rights', which must be defended by 'grabbing with two hands' (one to reform and open, while the other oppresses all kinds of criminal activities).[55] Opposed to him, even the demonstrators in Tiananmen Square in 1989 claimed to be representing the 'soul of China' (*guo hun*).

What the above draws our attention to is that we have to be extremely careful about the assumptions we make when talking about how ideas are globalised through communication. Whereas the illocutionary force of the international society metaphor for eighteenth-century jurists was to spread the Christian 'standard of civilisation',[56] in China the state/individual metaphor led to the institution of one-party dictatorship, justified in the eyes of the domestic audience in terms of national salvation, and in international arenas by what Vincent calls the 'morality of states'.[57]

The overall result of China's socialisation into international society, then, has been what some sociologists call a 'reverse discourse'.[58] Subjected to increasing international transactions, the Chinese state has not become a liberal-democracy, but has developed a political culture that is remarkably amenable to supporting totalitarian nationalism.[59] There is much evidence to suggest that the more recent combination of exposure to American culture and a commercialised publishing industry has continued to fuel this reverse discourse as much as it has led to any interest in liberal-democracy. The most extreme example of this is the best-selling collection of essays by young writers published under the title *China Can Say No*.[60] Here the authors quite clearly describe how their infatuation with all things American during the 1980s has given way to a bitter scepticism and demands for their government to take a stronger stand against Washington's attempts to bring about the 'peaceful evolution' of China. As for foreign travel and study, the same message emerges from a collection of essays under the title *Behind Demonizing China*, in which Chinese students who have studied in the United States express their dismay at what they see as the racism and China-phobia of the American mass media and China-watching community.[61]

If this brief survey indicates that inter-cultural communication is a creative process that does not lead to homogeneity, we must now address the question of how such diversity can be reconciled with the idea of political discussion between cultures.

GLOBALISATION OR GLOCALISATION?

If communication between cultures leads to diversity, it appears at first sight that only two possible positions can be taken on the problem of intercultural cooperation: either it is impossible due to the irreconcilable, and perhaps desirable, heterogeneity of the world's cultures; or it is possible through a process of homogenisation. However, sociologists looking more closely at this problem have increasingly found that this heterogeneity/homogeneity dichotomy is both sterile and does not provide an adequate explanation of how communication does actually take place without the need for homogeneity.[62] The sociologist Roland Robertson, for example, when examining what happens when religions have to compete with each other in a global context, observes that people are able to preserve a distinct tradition by relativising it in global terms. For religions, this can be seen in the ability to think in ecumenical ways. But it is also evident in other areas, such as the ability to think of local military-political issues in terms of 'world order', local economic issues in terms of 'world recession', citizenship in terms of 'human rights'.[63] Robertson has coined the term 'glocalisation', a term that originates from Japanese marketing jargon, for this ability to think about local issues in global terms.[64]

The great advantage of talking about glocalisation rather than globalisation is that it allows us to realise that the seemingly opposing trends of homogenisation and heterogenisation are in fact complementary and interpenetrative. This is because, although the end result of glocalisation is still the reproduction of the world as a single unit, it is still a world characterised by diverse interpretations of what that single unit is. What this means for global politics can be illustrated if we return to looking at the way in which the local collapse of the Qing dynasty was conceived in terms of a vocabulary and conventions generated during the process of socialisation into a global international society. Such a process cannot strictly be called 'Westernisation' because it involves the indigenisation of liberal ideas for the sake of solving China's specific problems, at the heart of which is the adaptation of the conventions of Chinese language. Because the resulting discourse is expressed in terms derived largely from addressing the problems created for China by international politics, though, it does at least refer to global problems in a way that could not have made sense before the expansion of international society.

This can be illustrated if we look at recent Chinese literature on international politics. A close reading of the essays in *China Can Say No*, for example, reveals that despite their nationalist xenophobia, even these authors are talking about glocal issues, such as how one can maintain one's identity in the modern world. In their case the problem is one of maintaining a 'Chinese' identity when confronted by hoardings boasting 'The Manhattan of the East' and 'China's Long Island'. This is a familiar problem throughout the world, although people in different cultures will arrive at different solutions. For at least one of these Chinese authors, the solution is not 'extreme nationalism', but some kind of state-supported preservation of native culture, along the lines of policies adopted in France and Canada.[65] Another author confronting this issue of how to respond to 'Western culture' can even find it in himself to explicitly argue that learning from outside China about a broad spectrum of social issues, from fashion to democratisation and economic reform, is as acceptable as 'going to a neighbour's house to make dumplings'.[66] Such views not only draw on foreign experience and address an issue of global concern, they are also ultimately compatible with the local ideological assumption that modernisation need not entail Westernisation.

When we turn to the work of Chinese academics, this tendency to think glocally is even more pronounced. It can be seen quite clearly, for example, in a collection of journal articles criticising Samuel Huntington's theory of the 'clash of civilizations'.[67] The introduction to this volume openly acknowledges that Chinese academics have moved from thinking in terms of geo-politics, through international political economy, towards understanding international problems in terms of world history and the future of civilisation. Partly stimulated by the 500th anniversary of Columbus's voyage to America and catalysed by the challenge posed by Samuel Huntington's civilisationalism, this has led to a lively debate and research agenda on globalisation.[68]

That the intention of the authors in this collection is to reject accusations that China will align with the Islamic states against the West, results in the use of a number of glocalising devices that provide an alternative discourse on Chinese nationalism. A good example of this is the tendency to relativise the crisis faced by Chinese culture by seeing it as one instance of a global problem.[69] That is to say, China's problem is not one of whether or not to Westernise, it is instead an instance of the universal problems of modernity that humanity faces as a whole. If this is the case, then it is not surprising that in rebutting Huntington, some of these authors explicitly welcome the questioning of the instrumental rationality of modernity that they find in Weber and Habermas, and the idea of a more human type of autonomous purposive rationality as offering a space for China to maintain some kind of cultural autonomy.[70] They see hope for diversity in the fact that Japan and Europe, after all, are similar to the US in economic terms, but very different in their cultures. Even within a Western state like the United States, immigrant groups maintain their separate identities, constituting, along with the women's movement, a possible starting point for the kind of cultural rebuilding that leads to diversity and offering a counterweight to the uniformity of modernity.[71]

When we look at the implications for nationalism of this de-centring of modernity, it is particularly interesting that it entails a movement away from the ethnocentricity of Chinese nationalist historiography. If 'the West' is decentred, there also needs to be an acceptance that China has absorbed other civilizations over the centuries, with the impact of Indian culture long preceding more recent calls to adopt 'science and democracy' from the West.[72] Turning to the meta-narrative of the expansion of international society, the destruction of civilisations by the West is acknowledged, but this does not represent the triumph of the liberal-democratic state, since a new situation has arisen of 'pluralism under a global consciousness' since World War Two.[73] The end of the Cold War acceleration can be understood from this perspective as due not to the triumph of the liberal-democratic state but to the pressure that has arisen as more and more nations have made social and economic development their priority, making them strong enough to resist the interference of the superpowers. Rather than the globalisation of liberalism then, the spread of the scientific-technological revolution, investment, information, manufacturing, trade, markets and international organisations is seen as leading to the creation of a material foundation upon which the American-European ethnocentricity of the Cold War can be replaced with peace, development, cultural pluralism and trans-cultural exchange. Unless, these authors believe, the West persists in seeing this as a challenge.[74]

What is most important to note about the glocalising processes that arise out of these responses to Huntington, though, is that in arguing against what is seen as an example of American neo-containment theory, they end up thinking in terms of local responses to global problems. A number of the authors are in fact grateful to Huntington for making them do this. It can be seen in the multiple-paths to modernity thesis, said to arise from the fact that humanity deals with common problems in different ways, varying according to whether a society is established on Confucian, Buddhist or Islamic foundations. It is most striking, though, in the proposition by one author that a holistic type of 'neo-rationalism' must develop under the pressure of finite resources and environmental crisis. The vision of a mutual interaction between the spirit of globalisation and the spirit of nativism (*bentu hua*) that is advocated here is as close to a statement of glocalisation that it is possible to get.[75]

The benefit of seeing the world in such fluid terms, for the Chinese of the era of 'opening and reform', is that they can identify themselves as being both Chinese and world citizens.[76] Nationalism is still there, and is firmly restated when the discussion focuses on sovereignty, but it is relativised in terms of solving global problems, and is seen as a double-edged sword.[77] They see that its nature varies depending upon the pace of development, with rash advances making it easy for extremism to arise and for the spirit of compromise to be lost, upon which all 'charters of civilisation' (such as the US Constitution) depend.[78]

Cultural globalisation is, then, something that people in China have been well aware of since Deng Xiaoping embarked on his 'opening and reforming' following the consolidation of his power in 1978. Since the 1980s there has been a flourishing of intellectual activity in the Chinese-reading world within which a complex debate on modernity and identity has played a prominent part.[79] If we are to grasp the political implications of such phenomena then it must be by seeing that people do make sense of state actions with reference both to the political cultures within which they act, and to global issues that threaten to affect their everyday lives. It follows from this that the more cultures are globally oriented, the more state actors will have to make sense of their policies in global terms.

An illustration of how this affects policy-making can best be given by looking at an area of activity that is pivotal for economic development in the global era, such as education. Ever since the Chinese Communist Party turned its attention to this area of reform in 1985, it has tried to strike a balance between surviving in a competitive world economy, while at the same time ensuring students are turned out who have 'ideals, morality, culture, discipline and ardent love for the socialist motherland and the tasks of socialism.'[80] The measures for achieving this may sound familiar, including giving schools and universities greater independence and making them more responsive to the demands of employers and the student market. When these have led to too much autonomy from the state, the immediate solution has been violence, as in the events of 1989. Yet in the longer term, the ideological solution has been even more vexing for those who would propound the idea of the globalisation of liberal-democracy. This is because, as the Cold War began to come to an end, the CCP turned to instilling 'patriotic education' into the population,

> letting students understand from a young age the splendid history and revolutionary tradition of the Chinese nation, understand the suffering of the Chinese nation over the past hundred years and the heroic struggle of the communist party in leading the people against imperialism and feudalism.[81]

After Tiananmen, this policy was combined with a campaign against 'bourgeois liberalisation' and 'peaceful evolution' instigated by the West, in which school children were taught to understand the meaning of exploitation, oppression, the imperialist invasion of China, and that 'without the CCP there is no New China; only socialism can save China; only with socialism can a Chinese way be developed.'[82]

If this is the context within which Chinese political literature is produced, it is not surprising that, from the most extreme xenophobic outburst to the more sophisticated work of academics in leading research institutions, local problems are situated in a context that is both global and nationalist. Such works no longer need to refer to Mao Zedong for confirmation of their views, but to Weber, Habermas and Liang Qichao, the latter of whom urged the Chinese to 'study as people of the country, study as people of the world'.[83] With there being more than 500 publishers in the People's Republic of China alone,[84] Chinese higher education institutes having Chinese-language web-pages, and Chinese-reading people all over the world having access to a wide variety of Chinese publications on the internet, including daily newspapers, the future development of Chinese identity has become a fervent topic of debate. Yet the current 'national study fever' (*guo xue re*) will have to result in an identity that makes sense both in terms of global problems, especially economic competition, and in terms of the specific problems of Chinese tradition. From this, at the very least, it might be concluded that the ability to think in local terms about global issues (and local issues in global terms) can take place without the institutions of liberal-democracy, and that it need not necessarily result in cultural homogenisation.

INTERNATIONAL RELATIONS AND THE LANGUAGE OF POLITICS

It has been argued here that arguments in defence of international politics lead irrevocably to the issue of inter-cultural communication. In fact, rather than international relations having to face the political, the post-Cold War situation appears to be one of *politics* having to confront the international, and ending up turning to culture in the process. This is because culture ends up being seen as the arena of social activity within which the liberal paradox needs to be resolved if political activity, in terms of discussion, is to take place at all. Although writers such as Held and Hirst and Thompson have not achieved a vision or explanation of how such discussion can take place in the resolution of global issues, they do at least leave us looking in the right direction. After all, the alternative to seeking a solution to the liberal paradox in the nature of cross-cultural understanding must be an increasing polarisation between those who see globalisation as desirable because it represents the triumph of the West,[85] and those who cling to heterogeneity out of a fear of the monotony and alienation of a homogeneous world.[86]

If this is the case, then it would seem that advances in areas such as linguistic and communication theory might contribute to an explanation of how global political activity is possible. Such a bringing together of disciplines might best be done through case studies on how communities enter international society, as has been attempted above with China. Here the process of socialisation has, of course, been traumatic, to say the least. It also appears to show that transactions between political communities do not necessarily lead to cultural homogenisation. If such transactions are to take place more with less disruption in future, then a central task for international relations should be to explore the conditions under which global issues can make sense in local terms through a process of true discussion between cultures.

It has been argued above that such discussion may be possible when states make sense of their actions with reference to globally oriented cultures, which is not the same as one homogeneous world culture. If this can be taken to mean that global communications need be no more of a threat to the existence of cultures than the telephone is to the personalities of individual people, then the liberal paradox need not be an obstacle to cooperation between states. Already, the Chinese case shows how a state with a relatively strong desire to protect its cultural autonomy is actually being glocalised in many ways. The same might be said of 'the West' where, although the spread of Chinese take-aways and kung-fu might not have been commented upon as much as that of the globalisation of McDonald's and Disney, the narratives of the 'Pacific century' and the 'tiger economies' have become central to political discourse.[87]

Precisely because so much of this kind of talk about globalisation is political mythology, however, there is a pressing need for a subject like international relations to engage in the discussion on how political activity across cultures takes place, and how it can take place more effectively. It has been pointed out recently that much can be contributed here by the ideas of writers on international relations such as Terry Nardin, who sees that politics in the international society of states is possible because it is a 'practical association' that shares no other common purposes than those of living together in peace and justice.[88] However, even such a view as this might need to be developed further by paying more attention to the nature of communication in international politics. Nardin's view, for example, that language itself is a kind of 'practical association' that has no intrinsic moral purposes,[89] needs to be qualified. As has been illustrated above, the intention behind the use of language can quite easily charge what are apparently neutral words in one culture with radical moral and political purposes in another. It is thus that a writer on problems of inter-cultural communication, like Robert Young, has to struggle to work towards a vision of what the language of a politics that is 'pragmatically-effective' might actually consist of.[90]

Although such an understanding of international politics might be a long way from providing a map for the realisation of cherished liberal-democratic principles, it does leave international relations an important role to play in developing a 'practical association' of states. If politics is increasingly turning to international relations in order to save its own relevance, and in the process ends up facing the paradox of different cultures, it seems to follow that the more international relations can take culture on board, the better will be its position to face the political in a global era. Of course, there should be little need to argue for the importance of developing our understanding of intercultural communication for a subject that is concerned with relationships between political communties. Moreover, as a discipline within which specialists on (and hopefully from) diverse cultures can communicate with each other under the umbrella of a broad theoretical debate, international relations has long provided an ideal arena within which to address such issues. Despite this, however, it is notable that some of the most significant attempts to look at language in international relations remain limited to understanding the nature of political language within the limits of English usage.[91] On the other hand, important work bringing linguistic theory to an area like Chinese politics tends to ignore international relations.[92]

Of course, in seeking a remedy to this problem, the 'delusions of etymology' and sterile debates on orientalism that reduce all politics to language should be avoided.[93] Yet such warnings should not be taken as prescriptions to avoid the more mundane demands of practical scholarship, where a number of obvious steps need to be taken. For example, the more that scholars are able to use non-European languages, the more the nature of language in international politics is likely to be understood. A thorough re-appraisal of the structures of academic institutions and curricula would also be desirable, with the study of non-European/American communities being properly integrated into core courses and the temptation to marginalise such subjects in specialist institutes resisted. If the mushrooming of Asia-Pacific Studies is anything to go by, the globalisation of the 'education market' might see that this takes place anyway. Who knows, perhaps university libraries can even be persuaded to acquire books written in the non-European languages of some of the world's largest and most powerful states?[94]

Ultimately, however, the main argument that has been made here is that if politics in a global era is to mean more than the tyranny of the markets, if 'engagement'[95] is to mean more than the homogenisation of modernity, or the creation of a world state, then the creativity involved in dialogue between cultures must be grasped and built upon. The more that research into different perspectives on globalisation is carried out, the more it will be realised how the essence of a truly global politics can be found in the potential that already exists in the dialogue between cultures that has been going on for many years. If civilisation comes into this at all, it is only in the sense of 'politics as a great and civilizing activity.'[96]

NOTES

1. See for example Paul Hirst and Grahame Thompson, *Globalization in Question* (Cambridge: Polity Press, 1996), and David Held, *Democracy and Global Order: From the Modern State to Cosmopolitan Governance* (Cambridge: Polity Press, 1995).
2. The classic application of liberal principles to international politics remains Hedley Bull, *The Anarchical Society: A Study of Order in World Politics* (London: Macmillan Press, 1977).
3. Bernard Crick, *In Defense of Politics* (Harmondsworth, Middlesex: Penguin, 1979).
4. Quoted by Crick, p. 17. This account of the state can be found in Aristotle, Book 2, *The Politics, The Politics and The Constitution of Athens* (Cambridge: Cambridge University Press, 1996), pp 30–61.
5. Crick, *op.cit.*, p. 33.
6. *Ibid.*, p. 18.
7. See, for example, William Wallace, 'Truth and power, monks and technocrats: theory and practice in international relations', *Review of International Studies* (Vol. 22, No. 3, July 1996), pp. 301–22.
8. James Shinn, 'Conditional Engagement with China', in James Shinn (ed.), *Weaving the Net: Conditional Engagement with China* (New York: Council on Foreign Relations Press, 1996), p. 7.
9. On 'spiritual civilisation' see Christopher Hughes, 'Globalisation and Nationalism: Squaring the Circle in Chinese International Relations Theory', *Millennium: Journal of International Studies* (Vol. 26., No. 1, Summer 1997), pp. 103–24.
10. Wallace, *op.cit.* p. 311.
11. Ken Booth, 'Discussion: A Reply to William Wallace', *Review of International Studies* (Vol. 23, No. 3, July 1997), pp. 371–77.
12. Hirst and Thompson, *op. cit.*
13. *Ibid.*
14. *Ibid.*, p. 173.
15. *Ibid.*, p. 180.
16. Shinn, *op.cit.*, p. 33.
17. Deng Xiaoping, 'Women you xinxin ba zhongguo de shiqing zuo de geng hao', ('We Have Confidence in Handling China's Affairs Even Better'), *Wen xuan (Selected Works)*, Vol 3. (Beijing: Renmin chubanshe, 1993), p. 325.
18. Held, *op. cit.*
19. Hu Yumin, 'UN's Role in a New World Order', *Beijing Review*, June 10–16, 1991, pp. 8–10.
20. Held, *op. cit.*, pp. 272–6.
21. *Ibid.*, p. 104.
22. *Ibid.*, p. 277.

Global Politics and the Problem of Culture   93

23. Samuel Huntington, 'The Clash of Civilizations?', *Foreign Affairs* (Vol. 72, No. 3, 1993), pp. 22–49.
24. 'Human Rights in China', *Beijing Review*, December 12–18, pp. 29–31.
25. Held, *op.cit.*, pp. 94–5. On this point see also Christopher Hughes, 'China and Liberalism Globalised', *Millennium: Journal of International Studies* (Vol. 24, No. 3, 1995), p. 442.
26. For a fascinating, first-hand insight into the predicament faced by dissidents in China when arguing for democracy in a political culture that puts the highest premiums on self-sacrifice and revolutionary fervour, see Liu Xiaobo, 'That Holy Word "Revolution"', in Jeffrey N. Wasserstrom and Elizabeth J. Perry (eds.), *Popular Protest and Political Culture in Modern China*, Second Edition (Boulder, CO.: Westview Press, 1994), pp. 309–24.
27. Held, *op. cit.*, p. 282.
28. *Ibid.*, pp. 60–1.
29. *Ibid.*, pp. 80–1.
30. *Ibid.*, p. 88.
31. *Ibid.*, p. 94. Perhaps it is rather ironic to point out here that by drawing on this meta-narrative of the post-colonial state, Held does in fact come very close to the views of Deng Xiaoping, who also likes to de-legitimise foreign interference in China's affairs by reminding us of historical events such as the invasion of China in 1900, which was carried out by forces from what are now the G7 states (minus Canada). Deng Xiaoping, 'Zhenxing zhonghua minzu' ('Stir Up the Chinese Nation'), *Wen Xuan* (*Selected Works*), Vol. 3., p. 358.
32. Hirst and Thompson, *op. cit.*, pp. 180–1.
33. Held, *op. cit.*, p. 284.
34. W. Dilthey, 'The Development of Hermeneutics', in *Dilthey: Selected Writings*, edited by H.P. Rickman (Cambridge: Cambridge University Press, 1976), pp 247–63. H.G. Gadamer, *Truth and Method* (London: Sheed and Ward, 1981).
35. Albrecht Neubert, 'Translation as Mediation', in Rainer Kolmel and Jerry Payne (eds.), *Babel: The Cultural and Linguistic Barriers Between Nations* (Aberdeen: Aberdeen University Press, 1989), pp. 5–12.
36. Ludwig Wittgenstein, *Philosophical Investigations*, trans. G.E.M. Anscombe (Oxford: Basil Blackwell, 1986), p. 6.
37. J.L. Austin, *How to Do Things With Words* (Oxford: Oxford University Press, 1975), pp. 94–108.
38. Jurgen Habermas, *The Theory of Communicative Action*, trans. Thomas McCarthy (Cambridge: Polity Press, 1987), pp. 6–20.
39. Zhang Yiwu interviewed in *Zhongguo da yuce* (*China Macro Survey*), Vol. 2, p. 687. The name of the book he is criticising is, *Duo yuan shehui zhong de wenhua piping* (no further details are given).
40. Habermas, *op. cit.*, pp. 25–7.

41. *Ibid.*, p. 96.
42. Edwin D. Dickinson, 'The Analogy Between Natural Persons and International Persons in the Law of Nations', *Yale Law Journal* (Vol. 26, No. 7, May 1917), pp. 564–591.
43. On the 'standard of civilization' see Gerrit W. Gong, *The Standard of 'Civilization' in International Society* (Oxford: Clarendon Press, 1984). A good overview of the relationship of 'civilization' to international law is provided in James Piscatori and Moorhead Wright, 'Cultural Diversity and International Law: Problems of Normative Order in International Relations', in Kenneth W. Thompson (ed.), *Community, Diversity, and a New World Order* (London: University Press of America, 1994), pp. 21–46.
44. This is a well known literary device, as when an author like James Joyce juxtaposes the words 'thoughts' and 'lice' in the sentence: 'His thoughts were lice born of the sweat of sloth'. Earl R. Mac Cormac, *A Cognitive Theory of Metaphor* (Cambridge, MA.: Massachusetts Institute of Technology, 1985), p. 23.
45. Liang Qichao, *Xin Min Shuo* (Taibei: Taiwan Zhonghua Shuju, 1978).
46. *Ibid.*, p. 1.
47. *Ibid.*, p. 6. For the concept of '*tian xia*' and the sinocentric world order within which it makes sense, see John K. Fairbank, *The Chinese World Order: Traditional China's Foreign Relations* (Cambridge, MA: Harvard University Press, 1968).
48. See especially Jan Nederveen Pieterse, 'Globalization as Hybridization', in Mike Featherstone, Scott Lash and Roland Robertson (eds.), *Global Modernities* (London: Sage Publications, 1995), pp. 45–68.
49. Kang Youwei, *Da tong shu (Book of Great Harmony)*, (Beijing: Guji chubanshe, 1956).
50. Mao Zedong, 'On the People's Democratic Dictatorship', *Selected Works of Mao Tse-tung*, Vol. IV (Beijing: Foreign Languages Press, 1975), p. 414.
51. Liang Qichao, *op.cit.*, p. 7.
52. *China Year Book 1934*, (edited by H.G.W. Woodhead, Shanghai), p. 385.
53. Shao-chuan Leng and Norman D. Palmer, *Sun Yat-sen and Communism* (London: Thamesand Hudson, 1961), p. 138.
54. This metaphor is especially prominent in appeals to the population of Taiwan for unification. See for example the 'Message to Compatriots in Taiwan from the Standing Committee of the Fifth National People's Congress', *Beijing Review*, 5 January 1979, pp. 16–17. For a more recent example the speech by President Jiang Zemin for the 1995 Lunar New Year, in *Renmin ribao (People's Daily*, overseas edition), 30 January 1995. For more on the origins of the Chinese discourse on race, see Frank Dikkoter, *The Discourse of Race in Modern China* (London: Hurst and Company, 1992).

55. Deng Xiaoping, 'Jianchi shehui zhuyi, fangzhi heping yanbian' ('Build Socialism, Oppose Peaceful Evolution'), *SW3*, pp. 344–6.

Deng Xiaoping, 'Zai Wuchang, Shenzhen, Zhuhai, Shanghai deng di de tanhua yaodian' (Essentials of Talks at Wuchang, Shenzhen, Zhuhai, Shanghai), *Wenxuan (Selected Works)*, Vol. 3, p. 378.

56. On the 'standard of civilisation' applied to China, Japan and Siam, see Gerrit W. Gong, *op.cit.*

57. R.J. Vincent, *Human Rights and International Relations* (Cambridge: Cambridge University Press, 1991), p. 123.

58. The idea of reverse discourse has been developed by Appiah from research on African reactions to imperialism. See Roland Robertson, 'Glocalization: Time-Space and Homogeneity-Heterogeneity', in Featherstone, Lash and Robertson (eds.), *op. cit.*, pp. 37–8.

59. Liu Xiaobo, 'That Holy Word "Revolution"', in Jeffrey N. Wasserstrom and Elizabeth J. Perry, (eds.) *Popular Protest and Political Culture in Modern China*, Second Edition (Boulder, CO.: Westview Press, 1994), pp. 309–24.

60. Song Qiang, Zhang Zangzang, Qiao Bian, *Zhongguo keyi shuo bu (China Can Say No)*, (Beijing: Zhonghua gongshang lianhe chubanshe, 1996). For two lengthy reviews of this work see Ju Gaujeng, 'Thoughts on Reading *The China That Can Say No*', and Christopher Hughes, 'A Western Scholar Looks At *The China That Can Say No*', *Sinorama*, Vol. 21, No. 11, November 1996.

61. Li Xiguang and Liu Kang (eds.), *Yaomohua zhongguo de beihou (Behind Demonizing China)*, (Beijing: Chinese Academy of Social Sciences, 1996).

62. On the nature of this dichotomy, see Robertson, 'Glocalization', *op. cit.*, p. 27.

63. Malcolm Waters, *Globalization* (London: Routledge, 1995), p. 42.

64. Robertson, 'Glocalization', *op.cit.*, pp. 25–44.

65. Tang Zhengyu, 'Ezhi fan er jiang zhongguo daoru le geng da de guoji hezuo huanjing' ('Containment will lead China to a Situation of Even Greater International Cooperation'), *Zhongguo keyi shuo bu*, (*China Can Say No*), (Beijing: Zhonghua gongshang lianhe chubanshe, 1996), pp. 198–9.

66. Gu Qingsheng, 'Bie ba ziji gao de hen zibei' ('Do Not Put Denigrate Ourselves'), *Zhongguo keyi shuo bu*, p. 279.

67. Wang Jisi (ed.), *Wenming yu guoji zhengzhi: zhongguo xueje ping Hengtingtun de wenming chongtu lun*, (Culture and International Politics: Chinese Scholars Criticise Huntington's Theory of the Clash of Civilizations), (Shanghai: Renmin chubanshe, 1995). Hereafter referred to as *Wenming*.

68. Chen Lemin, 'Tuokuan guoji zhengzhi yanjiu de lingyu' ('Broaden the Area of Research in International Politics') in *ibid.*, p. 3.

69. Robertson points out that this device is also used in Japan. See 'Glocalization', *op. cit.*, p. 28.

70. Jiang Yihua, 'Lun dongya xiandaihua jincheng zhong de xin lixing zhuyi wenhua', ('On the New-Idealist Culture in the Process of East-Asian Modernization') in *Wenming, op. cit.*, p. 263. Zhang Rulun, 'Wenhua de chongtu haishi wenhua de kunjing' ('Cultural Conflict or Cultural Predicament') in *Wenming, op. cit.*, p. 319.
71. Zhang Rulun, *op. cit.*, p. 320.
72. Tang Yijie, 'Ping Hengtingtun de "wenming de chongtu?"' ('A Critique of Huntington's "Clash of Civilizations?"') in *Wenming, op. cit.*, pp. 251–3.
73. Tang Yijie, *op. cit.* p. 256.
74. Jiang Yihua, *op. cit.* pp. 275–6. Feng Tianyu, '"Wenming chongtu jueding lun" bianxi' ('Analysis of the "Theory of the Determinism of the Clash of Civilizations"'), in *Wenming, op. cit.*, p. 303.
75. Jiang Yihua, 'Lun dongya xiandaihua jincheng zhong de xin lixing zhuyi wenhua', ('On the New-Idealist Culture in the Process of East-Asian Modernisation'), in *Wenming, ibid.*, pp. 260–275.
76. Feng Tianyu, *Wenming, ibid.*, p. 306.
77. Wang Jisi, '"Wenming chongtu" lun zhan pingshu' (Review of the Debate on "Clash of Civilisations" Theory) in *Wenming, ibid.*, p. 37. The same point is made in Feng Lin (ed.), *Zhongguo da yuce (China Macro Forecast)*, (Beijing: Gaige chubanshe, 1996), Vol. 2, p. 658.
78. Xu Guoqi, 'Hengtingtun ji qi "wenming chongtu" lun' ('Huntington and His Theory of the "Clash of Civilizations"') in *Wenming, ibid.*, p. 93.
79. Aside from the political issues that this raises for China, those who wish to move away from the ethnocentricity of globalisation theories should also note that this debate is being globalised through the Chinese speaking world, including Taiwan, Hong Kong, Singapore, and the Chinese diaspora. A good survey of this can be found in Lin Tongqi, Henry Rosemont Jr., and Roger T. Ames, 'Chinese Philosophy: A Philosophical Essay on the "State of the Art"', *The Journal of Asian Studies* (Vol. 54, No. 3, August 1995), pp. 727–58. See also Tu Wei-ming, 'Cultural China: The Periphery as the Center', in Tu Wei-ming (ed.), *The Living Tree: The Changing Meaning of Being Chinese Today* (California: Stanford University Press, 1994), pp. 1–34.
80. The key reformist document for education was the *'Zhonggong zhongyang guanyu jiaoyu tizhi gaige de jueding'* ('Resolution of the Central Committee of the CCP on Reform of the Education System'), 27 May 1985, published in *People's Daily*, 29 May 1985.
81. *'Zhonggong zhongyang guanyu gaige he jiaqiang zhong xiao xue deyu gongzuo de tongzhi'*, ('Memorandum of the Central Committee of the CCP Regarding Reforming and Strengthening Moral Education Work in Middle and Junior Schools'), 25 December 1988.
82. *'Guojia jiaoyu weiyuan hui guanyu jin yi bu jiaqiang zhong xiao xue deyu gongzuo de yijian'* ('Views of the National Education Commission Concerning A Step Forward in Strengthening Moral Education Work in

Middle and Junior Schools'), 13 April 1990.
83. *Ibid.*
84. There were 562 in 1995, which had grown from a figure of 105 in 1978. *Chuban ren (Publisher)*, (August 1997, Taibei), p. 174.
85. Francis Fukuyama, *The End of History and the Last Man* (London: Penguin, 1992), is the most notable example of this. As has been shown above, though, elements of such thinking also pervade some of the best intentioned works on globalisation.
86. For a recent attack on the desirability of a homogenous global culture, see Anthony D. Smith, *Nations and Nationalism in a Global Era* (Cambridge: Polity Press, 1996).
87. See, for example, Tony Blair, *New Britain: My Vision of a Young Country*, (London: Fourth Estate, 1996), pp. 98,110, 119, 203–4, 260, 296.
88. Terry Nardin, *Law, Morality and the Relations of States* (Princeton: Princeton University Press, 1983), and the discussion in Chris Brown, *Understanding International Relations* (Basingstoke: Macmillan Press, 1992), p. 54.
89. Nardin, *op. cit.*, p. 6.
90. Robert Young, *Intercultural Communication: Pragmatics, Genealogy, Deconstruction* (Clevedon: Multilingual Matters Ltd, 1996), p. 208.
91. See, for example, Francis A. Beer and Robert Hariman, *Post-Realism: The Rhetorical Turn in International Relations* (East Lansing, MI.: Michigan State University Press, 1996).
92. See, for example, Michael Schoenhals, *Doing Things with Words in Chinese Politics* (Berkeley, CA.: University of California, 1992).
93. Fred Halliday, 'Orientalism and Its Critics', in *Islam & the Myth of Confrontation: Religion and Politics in the Middle East* (London: I.B. Tauris, 1995), pp. 195–217.
94. Books from the PRC can be obtained using credit card on http://www.modernbooks.com.
95. For an overview of US views on 'engaging' China, see Shinn (ed.), *op. cit.*
96. Crick, *op. cit.*, p. 15.

5. Overturning Globalisation: Resisting the Teleological, Reclaiming 'the Political'[1]
L. Amoore, R. Dodgson, B. Gills, P. Langley, D. Marshall, and I. Watson[2]

FIRST ORDER QUESTIONS: DO WE ACCEPT 'GLOBALISATION'?

The term 'globalisation' has served as an arresting metaphor to provide explanation, meaning, and understanding of the nature of contemporary capitalism, though not all of the processes that currently come under the rubric of globalisation are new.[3] It is meant to suggest a number of analytically distinct phenomena and developments within the international system, while combining them into a single overarching process of change. Considerable attention centres on the application of new (often information based) technologies to the production process, and parallel changes in management, organisation, and communications at corporate, societal, and state levels.

Moreover, the current vogue has been to establish globalisation as both epoch and epistemology. Its ideational foundation is rooted in notions of 'progress' and perpetual change. Neoliberalism seemingly triumphs in the intellectual firmament, as modernisation theory is resuscitated to forecast convergence of economic and political systems across the globe due to the inexorable processes of capitalisation/capital accumulation. In the words of one recent critical commentator, however, '[i]n reality, globalisation is to the world economy what monetarism is to the domestic economy. It represents the final triumph of capital over labour...'[4] The OECD promotes a Panglossian view of globalisation as a liberal utopia giving 'all countries the possibility of participating in world development and all consumers the assurance of benefitting from increasingly vigorous competition between producers.'[5] On the other side of the debate, as eminent a figure as J.K. Galbraith sums up the trends of the past fifteen years as 'The Uncertain Miracle' and warns of 'the possibility of a depressive equilibrium as regards unemployment.[6] Ralf Dahrendorf worries '[h]ow can the affluent societies of the world retain their wealth, freedom, and social cohesion in the face of the destructive pressures of economic globalisation?'[7] Other economists worry that the trade liberalisation and financial deregulation of the market dominated 1980s and 1990s have been associated with deflationary economic outcomes, while countries importing huge amounts of footloose capital can be destabilised rather than developed. The upshot is that rampant neo-liberal economic globalisation may not result in a new global utopia, but rather in a global dystopia.

Even though a number of scholars critique the triumphalist strain within globalisation discourse[8] they nevertheless seem to accept the basic assertion that contemporary capitalism has entered a new phase; that is, globalisation as epistemology leads to globalisation as epoch. Thus a tendency to a univocal discourse (in this respect) has become a problem, and should be re-examined. For example, neo-classicists take the view that finance, culture, markets, and production are now so sufficiently interlinked as to constitute a change in global capitalism of abiding significance, making the retreat of the state inevitable and irreversible. Postmodernists and post-industrialists tend to mark the epochal shift in capitalism as occurring in the wake of significant 1970s adjustments. Neo-Schumpeterians echo these views by emphasising the impact of technological change on the industrial base of core economies and thus on global capitalism. Regulationists and post-Fordists place emphasis on changes occurring in core firms, i.e. 'leaner' organisation and management and 'flexible' production techniques. Neo-Gramscian scholars establish a changing order by referring to the transnational policy influence of an epistemic community of business elites, imbricated with a network of officials in governments and in various multilateral institutions.

The effect of all this analysis is too often, in our view, to create and sustain an assumption that globalisation is manifestly obvious and inexorable. It is precisely this fundamental assumption that we wish to challenge. Globalisation is a contested concept, not a received theory. In our view, it is a serious analytical, as well as political mistake to begin from the assumption 'globalisation *is*.' This raises the question of whether there is some 'single logic' at work in the processes of globalisation, and our attitude to such a concept. The answer to this question depends on what is meant by 'logic'. If the logic of globalisation is taken to mean something external to society and the state and inexorable, then we firmly reject this notion of a single logic. If on the other hand, a single logic is conceptualised as a set of common pressures or trends that we are all becoming increasingly bound up in, even though the patterns of responses are uneven, then we can provisionally accept such a concept.[9] However, it may be clarifying to talk about logic in terms of processes of capital accumulation, for example, as opposed to *the* logic of capital accumulation.

We arrive at a formulation of the problem whereby globalisation as over-determined 'reality', that is, as something 'external', has to be rejected, yet certain processes which currently come under the rubric of globalisation and have real and felt damaging effects on people must be recognised and 'resisted', a process which begins by 'demystifying' them. Therefore, we insist that it is necessary to ask where these realities come from, that is, does globalisation originate from the ideological sphere? Thus, globalisation has to be resisted both as ideology,[10] and therefore in a struggle against intellectual opponents through elaboration of an alternative political economy, as well as at the level of the practical and political, i.e. via 'concrete strategies of resistance'.

We seek in the remainder of this chapter to highlight some of the problems associated with deploying the concept globalisation in our research agenda. This is done by questioning assumptions made about the presumed definition of globalisation and its relationship with the state, civil society, and social movements (old and new).

WHAT IS GLOBALISATION? THE DEFINITIONAL PROBLEM

Most scholars seem to agree that globalisation encompasses a broad range of material and non-material aspects of production, distribution, management, finance, information and communications technologies, and capital accumulation. The most visible effects or processes of globalisation seem to be: the increase in the speed and flow/flight of capital in money form; the expansion of offshore financial markets; the advance of computer driven technologies; and the renewed impetus towards regionalisation. The choice of the root 'global' (implying a totality) and its transformation into an action-process verb (-isation), seems to imbue the concept with a special meaning and social power.

When globalisation is narrowed to the neo-liberal project of economic globalisation, then the outlines of a definition become clearer. In essence, we would contend that neo-liberal economic globalisation has four defining characteristics: (1) to protect the interests of capital and expand the process of capital accumulation – if this is viewed as occurring within and because of a structural crisis in capitalism or a long term economic stagnation, then neo-liberal economic globalisation is essentially a strategy of 'crisis management' or 'stabilisation'; (2) the tendency towards homogenisation of state policies and even state forms towards the end of protecting capital and expanding the process of capital accumulation, via a new economic orthodoxy, i.e. 'market ideology' – wherein even the state itself becomes subject to marketisation while simultaneously being deployed instrumentally on behalf of capital; (3) the addition and expansion of a layer of transnationalised institutional authority 'above the states' – which has the aim and purpose of penetrating states and re-articulating them to the purposes of global capital accumulation; and (4) the exclusion of dissident social forces from the arena of state policy making – in order to insulate the new neo-liberal state form against the societies over which they preside and in order to facilitate the 'socialisation of risk' on behalf of the interests of capital.

Nevertheless, the debate on globalisation is obscured by a pervasive conceptual fuzziness surrounding the term itself. The definition of globalisation exists in a wide range of permutations, with little consensus between the different approaches. There is certainly a set of 'clusters' of definition, according to the perspective or issue area from which the definition emanates. A crude typology of such definitional clusters might include the following categories: economic processes; political processes; world culture processes; and global civil society processes. As a recent discussant in the IPE debate on 'globalisation' on the internet commented 'Within each cluster, there is a substantial degree of variation in the degree of theoretical self-consciousness of usages of the term, the facts of globalisation emphasized in the definition, and the degree to which "globalisation" is considered to be a multidimensional process.'[11]

If we are to identify a common element across most of the prevailing approaches, then the notion of an 'epochal shift' does appear to be key. This idea consists of the notion that all societies, stimulated by the forces of global change, are taking on 'new' forms. These new configurations are characterised variously as, for example, 'post-Fordism',[12] postmodernity[13] and post-industrialism.[14] There is, then, a general notion of reorganisation 'after' something, though little agreement exists on exactly what the important crisis-ridden event(s) was or were. Nevertheless, the idea of profound discontinuity seems to prevail over the idea of fundamental continuity.

Although there is a certain attractiveness in avoiding precise definition, and of staking out a 'break' in history and declaring a 'cross-roads' in historical capitalism, these are problematic from a 'social science' point of view. The reification of a 'new phase' and according it iconic status is a device which opens the gates to a conflation of theories. Misshapen fragments of ideas become integrated into a generalised overarching explanation. Thus, 'globalisation' becomes a 'horse for every course'. Diverse perspectives are conflated in order to provide evidence of wholesale change. 'Globalisation' tends to become a 'totalising notion', and at the very least there is a strong tendency to eclecticism. For example, Castells and Hall[15] seek to explain changes in cities and regions through combining the dynamics of technological revolution, the formation of a global economy, and the emergence of the 'information age'. Bob Jessop uses an eclectic mix of post-Fordist thought, combining elements from the regulation school and the neo-Schumpeterians.[16] The problem of conflation of theories, which is a more serious matter than mere eclecticism, also manifests itself in terms of the empirical evidence deployed in analyses of globalisation processes. Empirical evidence is commonly drawn from specific cases or 'ideal type' models which are then 'generalised up' to explain the whole picture; a kind of positivistic holism. For example, the 'industrial districts' of Third Italy and Baden Wurtemburg are offered as models for the post-industrial production system.[17] The result of conceptual conflation is broadly twofold; on the one hand for those seeking to manage the contemporary situation, the conceptual incoherence breeds simplicity in a complex world.[18] A new grand narrative may be emerging from disparate fragments of global analysis. On the other hand, for those unable to untangle the conflated strands of the debate, it breeds genuine confusion.

THE TELEOLOGY OF 'GLOBALISATION'

In seeking to unravel some of these ideational strands, it is useful to draw out what we consider to be the flawed assumptions underlying the dominant discourse of globalisation. The five points we briefly review below are all aspects of the teleology of globalisation and make clear its apolitical impetus. Social agency in the form of the state, social movements, and organised labour are all basic to forms of politics we currently understand and practice, yet the teleology of globalisation tends to produce the idea of the 'death of politics' as well as the demise of the nation-state.

1. *Technological change* is presented as the driving force of globalisation, i.e. changes in science, technology and production methods essentially determine the future for workers, managers, the state, and their inter-relationships. The introduction of new technologies heralds a transitional crisis as social and political institutions strive to catch-up, but via a pre-determined set of forms. Globalisation, then, is conceived in a teleological manner. Technological change is an 'iron cage', and 'everyone is doing it'. These processes are presumably beyond our control.

2. Globalisation is framed in the *essentialistic*. Changes in economics, events such as the oil crises, the break-up of Bretton Woods, and other macro-structural causes are variously interpreted as compelling societies to 'make the leap' to 'globalisation' as if there were no other alternative. The precise contours are unclear, however. Are we jumping from Fordism to post-Fordism, from modernity to postmodernity, from the industrial to the post-industrial, or all three perhaps? What is omitted in the essentialistic reading of globalisation is the notion of the historically and socially embedded conditions and relations, which themselves condition the outcome of complex socially constructed 'events' such as 'globalisation'.

3. There is a strong emphasis, from many perspectives, on the notion of *convergence*, which is the idea that from divergent

starting points and diverse institutional bases, societies become increasingly alike. For example, there is supposedly convergence at the level of the nation-state as its borders are transcended by the globalisation of finance, production and trade.[19] Technological advances in financial systems are said to herald the arrival of 'quicksilver capital'[20], and new structural forces in the world economy.[21] The rapid maturation of technologies is equated with the absolute necessity for firms to 'go global'.[22] Culture is said to be undergoing homogenisation as technology breaks down traditional cultural territories.[23] The global competition imperative is overwhelmingly adopted as a business mantra, provoking debates regarding the 'right' path for the twenty-first century organisation of the production and labour processes,[24] and indeed the right path for nations (i.e. states) to follow.[25] But is this ahistorical abstraction any more than (Almondian) modernisation theory 'gone berserk'? We have moved from the earlier idea of convergence between East and West towards a state-managed social democracy, to a presumed convergence between North, South, East and West, towards a post-statist neo-liberal market society wherein the state and labour have lost power to corporations.

4. Globalisation is presented *instrumentally*. There is a tendency to simplify the description of change in order to prescribe a set of formulae to manage change. Debate surrounding global change finds academic analysis cutting across business management guides and political campaigns. Academic commentary, policy discourse, and corporate strategy become ever more closely intertwined.[26] The result is a preoccupation with both political 'strategic' crisis management, and 'strategic' business management. However, there is a political deception at the root of this instrumentality. The political choice to liberalise and lower barriers to trade and international money in the late 1970s and during the 1980s involved a choice by governments of how to present this change to the public. Governments normally chose not to argue that the presumed benefits of liberalisation (higher aggregate wealth) would outweigh the costs (lower wages and

higher unemployment in industries facing new competition) since, '[t]o argue that way would only have drawn attention to the costs. Better to say that changes in the world economy left no choice but to liberalise.'[27] At the corporate/media level the concept is instrumentally presented to society in the advertising campaigns for everything from credit cards to soft drinks. We are encouraged to equip ourselves with these essential products and services for life in the 'global village'.

4. Globalisation is presented as a *benign* process. Social conflict is posited as being confined to the adjustment phase, with new practices promising 'worker empowerment', the formation of 'new solidarities', and even the return of a craft-based 'yeoman democracy'.[28] As the institutions of the old order give way to new structures, conflict is viewed as a temporary by-product of restructuring. Globalisation becomes a harmless automatic process that transcends 'the political', perhaps once and for all. As the director of the OECD implied in an interview in *Le Monde* published in 1992, entitled 'globalisation is here to stay'[29], our task is to accept and adjust. Even on the left of the academic spectrum, a number of authors make the mistake of assuming that 'globalisation is here to stay', even if endorsing a variety of presumed national or regional regulatory responses. Such 'tinkering at the margins' of globalisation, however, has the unintended effect of helping the globalisation project to consolidate itself. This approach is therefore conservative (with a small c), in the sense of conserving the main theses of globalisation, instead of making a determined effort to 'overturn globalisation'.

So far the questions posed contra globalisation have on the whole still not been sufficiently framed in the explicitly *political*. Nevertheless, globalisation is now being challenged by a more critical question-raising and empirical research agenda. The clarification of the definition of globalisation and the 'testing' of the concept has recently become a first order issue in research in this field. To what extent is globalisation distinguishable from interdependence?[30] How important is the distinction between multi-national corporation and trans-national corporation?[31] We are now seeing a rise in academic interest in persistent elements of diversity, divergence, conflict and difference. The 'new institutional' economics and economic sociology of recent years raises pertinent questions as to the 'embeddedness' of social practices and the social and historical nature of change.[32] Simultaneously we are seeing an increased emphasis on the study of 'national trajectories' in, for example, welfare and industrial restructuring.[33] Rigorous testing in terms of empirical examination and the study of 'bearers of change' has become more visible in recent years.[34] The rise in critiques based around the use of 'myth' and ideology support the general trend towards questioning, stripping the concept to its bare essence to expose its political dangers.[35] Attempts to expose the dangers of the concept have also been presented to the public through the media.[36]

GLOBALISATION AND THE STATE: RETREAT OR RETURN?

Many globalisation theorists suggest a profound disjuncture between the persistence of territorial sovereignty as the organising principal of international politics and the increasingly global structure of finance, production, society and culture. This position has two fundamental consequences. First, traditional paradigms which have taken the territorially bounded nation-state as the central unit of analysis are rendered redundant. Globalisation becomes epistemology as the study of human action and interaction is perceived to require a transnational mode of enquiry.[37] Second, the presumed erosion of national economies and societies raises a question mark over the potential political power of the state in the coming global era. Both 'inside' and 'outside' territorial borders, the power of governments and institutions to shape the economic and social environment in which they find themselves appears to have waned. Such consequences demand that the relationship between the state and the processes of global restructuring be central to the globalisation debate. To this end, it is possible to identify the emergence of three broad schools of thought.

The currently dominant conceptualisation of the relationship between globalisation and the state views transnationalisation as seriously undermining the basis of the nation-state as a territorially bounded economic, political and social unit. The structure of the wholesale financial system is held to have been transformed, from primarily national with some transnational linkages to largely global with some national differences.[38] Similarly, national corporations that in the past firmly associated themselves with their 'home' economy are believed to have been replaced by genuinely transnational corporations (TNCs).[39] Once relatively cohesive national societies are now held to be undermined by global cultural flows carried across borders by technological and media infrastructures.[40]

Consequently, state authority over economy and society is viewed as undergoing a period of diffusion; leaking 'upwards' to inter-state institutions, global social movements, and TNCs; shifting 'sideways' to global financial markets and the more powerful states; and descending 'downwards' to quasi-public regional and local institutions.[41] Further, it is suggested that state structures, institutions and policies converge around a complex and contrary melding of neo-liberal rhetoric and the competitive provision of incentives for transnationally operating capital. States lose their economic policy making sovereignty, becoming locked in to the financially orthodox policies of low inflation and 'strong' and 'sound' public finance by the power of global financial markets.[42] The nature of the competitive struggle between states has shifted from conflict for control of territory and wealth-creating resources to the competitive provision of human and physical infrastructures in order to attract transnationally operating capital,[43] with the concomitant decline in the significance of defence and foreign policy and increase in the importance of industrial and trade policy.[44] Fundamentally, the result of the dominant conceptualisation is an understanding of the processes of global restructuring in which the state is viewed as merely responding to market, cultural and technological changes occurring above and beyond the state itself.

A direct challenge to this commanding orthodoxy is the claim that the 'decline of the state' in the face of the forces of globalisation has been exaggerated. The basis for this view is the recognition that the foundations of the global economy remain largely national in character.[45] National capitalisms do not converge around a single form but persist, particularly with reference to alternative state-finance-industry relations.[46] Nationally specific state structures and institutions continue to interact with the international political economy ensuring that national development trajectories are sustained.[47] Analysis of key wholesale financial indicators such as interest rates, securities trading, and savings and investment illustrates that contrary to the popular vision of 'seamless' global markets, most economic variables remain firmly national in orientation.[48] Similarly, examination of the ownership, management and organisation of the world's largest corporations suggests that they maintain established associations with their national bases, and therefore are best characterised as 'multinational'.[49] As a consequence, states are held to continue to exercise considerable authority in the management of the national economy.[50] Similarly, for the 'regulation school' the system of national monetary management persists as it is a crucial component of the 'mode of regulation', that is the institutional milieu which underpins capital accumulation.[51] Ultimately, this understanding of the relationship between the state and globalisation suggests that the processes of global restructuring have been driven by the interaction between contending national capitalisms. As Zysman asserts,

> National developments have, then, driven changes in the global economy; even more than a so-called "globalisation" has driven national evolutions. It is the success of particular countries, rather than some unfolding of a singular global market logic based on more and faster transactions, that has forced adaptations.[52]

The argument that the restructuring of the contemporary era has and continues to be determined by the distribution of power within and between the major nation-states and their rival capitalisms leads to changes being characterised as 'internationalisation' and not globalisation.[53]

A third position suggests that the usual understanding of a dichotomy between the state and globalisation is an illusion, as the processes of global restructuring are largely embedded within state structures and institutions, politically contingent on state policies and actions, and primarily about the reorganisation of the state.[54] The nature of state intervention may have changed, but the state has not necessarily diminished in its significance to contemporary capitalism. Some parallels can be drawn here to Polanyi's approach to the so-called laissez-faire state associated with the rise of industrial capitalism, in that the separation of economy from polity is recognised to require the exercise of state power.[55] In its minimalist form, this view identifies the continuing role of the state as the ultimate guarantor of the rights of capital whether national or foreign.[56] More broadly, however, the 'internal' restructuring of the relationship between the state and national economy/society and the 'transnationalisation' of the state itself, as national systems of regulation are replaced by global systems of regulation, are both identified as fundamental to globalisation.[57] A new layer of politics above the state is highlighted, based around increased national regulatory co-operation and new institutional forums.[58] In essence, this interpretation places the state at the centre of analysis of globalisation, viewing the processes of change as shaped by the restructuring of political power.

Taken together, the alternative schools of thought outlined above suggest that, contrary to the dominant view, globalisation is a set of multi-dimensional processes in which the state is not transcended, but rather is of fundamental significance. In short, the immediate sources of globalisation are to be found precisely in the 'political' domain. The research agenda which follows from this suggests the need to challenge the dominant techno-rational interpretation by asking how and why state institutions and policies came to be at the heart of globalisation. We are able to move beyond the dominant discourse, which suggests that there is no alternative to the neo-liberal state form, to show that the processes of restructuring are politically contingent on the predominance of neo-liberal forces and policies at the level of state politics.

The position that the national state is still the most important site of social change or still capable of controlling a national economic development trajectory is necessary, but not sufficient to 'overturn globalisation'.[59] Likewise, reclaiming the 'national economy' and state manoeuverability are positive steps towards recognising social contestation and the centrality of the *political* process, but insofar as they remain bound in these statist frameworks they remain limited. We must move beyond 'technicist' arguments on the continued role and efficacy of the state. They do not yet constitute an unambiguous progressive counter-position to neo-liberal economic globalisation. Once it has been shown that globalisation and the immediate responses to it are indeed squarely in the realm of 'the political', this can form the basis of expanding the domain of the political further to press for *social* as opposed to narrowly economic concerns.

Such a counter-hegemonic position must concentrate on the question of *social reform* and the changing relationship between civil society and the state or between social forces and state power. Civil society has recently become the subject of much academic interest. This new interest has concentrated around the idea of civil society being seen as the source of a socio-political challenge to the state, capitalism, and the inter-state system.[60] Civil society therefore can no longer simply be seen as an agglomeration of institutions and social relations which are separate from the state. Indeed, following Gramsci on this point, it is our view that:

> civil society is the sphere of class struggles and of popular democratic struggles. Thus, civil society is the sphere in which a dominant social group organises consent and hegemony. It is also the sphere where the subordinate social groups may organise their opposition and construct an alternative hegemony – counter-hegemony.[61]

The caution over the uncritical use of statist frameworks is particularly the case when evidence from East Asian developmental states is deployed in defense of the national economy/state manoeuverability position, given that most of these 'models' were authoritarian. The 'return of the state' position, though a welcome antidote to the 'retreat of the state' position, is incomplete without further analysis of and focus upon the recomposition of social forces and their need and capacity to alter or channel the political direction of the state and the economy or, in Polanyian terms of the 'double movement', the prospect of society regaining control in response to the destabilising effects of the unregulated market.

GLOBALISATION, SOCIAL AGENCY, AND RESISTANCE

The first order (research and practical) question has to be: how do we frame resistance to 'globalisation', as distinct from resistance to capitalism, or to imperialism? Far from being an age free of social conflict, 'resistance' seems to be on the rise around the world, and taking myriad forms and directions. It is possible that we have already seen the first mass political 'revolts' against globalisation. In the developed world, the three weeks of strikes and protests in France in December 1995[62] were directed against 'restructuring' (albeit in a context of the politics of European integration, i.e. regionalisation), and in the developing world, the armed revolt of the Zapatistas pitted the peasantry of Chiapas directly against the forces of economic neo-liberalism and NAFTA, and inspired a new indigenous peoples movement and the national democratic opposition.[63]

The forms of resistance which are emerging are indeed diverse, and this means that differentiation is needed between forms of resistance, since not all are moving in the same direction or could be categorised as necessarily 'progressive' or acting directly in opposition to forces of neo-liberal economic globalisation. However, this new politics is no longer confined exclusively to 'national politics' in the traditional sense. Included amongst the new 'sites' of resistance are: local communities; indigenous peoples' communities; urban community organisations; organised labour; old and 'new' social movements. Moreover, these emerging forms of resistance act in and across different spatial scales; for example, the local, the national, the regional and the global. Thus, for example, the Zapatista movement can be seen as an organisation of indigenous peoples which is based in and operates largely on the local level, yet has a larger appeal and a real political impact at national level, and even some 'influence' at regional and global level, at least in its outreach for solidarity and its inspirational 'demonstration effect'. SANCO[64], the South African National Civic Organisation, is an example of an urban based movement which operates primarily on a national level, representing a new relationship between civil society and the re-articulated South African state. The global women's health movement can be seen as an example of a social movement which operates across the local, national, regional and the global level,[65] representing a new global social movement committed to emancipation. The issues on which these resistance groups campaign may seem diverse and unrelated in nature. However, the link between indigenous people from Chiapas demanding land reform and national social justice[66] and a coalition of churches, NGOs and social movements challenging the World Bank's structural adjustment programmes in Africa[67] or the campaign to establish new international standards of labour rights[68] is that the issue on which they are campaigning can be understood as a product of the same phenomena – market-driven economic globalisation and its destabilising effects on society.

At present, there is both optimism and pessimism about the ability of emerging resistance movements to mount an effective challenge to neo-liberal economic globalisation. Those authors who are optimistic about the future of the contemporary resistance movements to globalisation point to the broad popular appeal of these resistance movements[69] and to the struggles won by grassroots resistance movements against economic restructuring in countries such as India, Mexico, Nepal and Costa Rica.[70] From this optimistic perspective, diverse resistance movements can be seen as representing an effective challenge 'from below' to economic globalisation. Other authors argue that to be successful, the challenge to globalisation must now be made on a global level. This point is made by Christopher Chase-Dunn[71] and also by Robert Cox who comments that, 'the counter force to capitalist globalisation will also be global.'[72] John Cavanagh[73] and Richard Falk[74] also recognise the emergence of global social forces' challenges to neo-liberal globalisation. The 'Fifty Years is Enough' network, for example, campaigns for the abolition of the World Bank.[75] New efforts to organise a global labour movement challenging the ethos of flexibilisation is another example of global challenge to global economic forces.[76] Commenting on this, Chase-Dunn has argued that 'internationalism has finally become not only desirable but necessary.' However, '[t]his does not mean that local, regional and national-level struggles are irrelevant. They are just as relevant as they always have been. But they need to also have a global strategy and global-level cooperation lest they be isolated and defeated.'[77] In a similar vein, activists and intellectuals engaged in the struggles of women workers recognise the need for new strategies of resistance, including building stronger international alliances.[78]

In opposition to the above optimistic view it can be argued that historically we have seen instances of communal (e.g. Luddism), national-popular (e.g. the Cuban Revolution) and international (e.g. the Communist Internationals) resistance to capitalism. All of these previous attempts at resistance have failed to halt the advance of capitalism around the globe. Why then should we expect contemporary forms of resistance to be any more successful in securing a victory against neo-liberal globalisation? Drainville muses that 'the new internationalism... seems to involve little beyond a vague humanism, a concern for the victims of the world economy' while 'the new internationalism of social movements... is a mutable and ever-changing collection of narrowly focused social movements that are continually reminded of their transitory nature by the unrelenting restructuring of production in the world-economy.'[79] Wagar is particularly scornful about the ability of emerging resistance movements to mount an effective challenge to globalisation. As Wagar comments, 'I would caution... against investing too much hope in the nominally or apparently anti-systemic movements visible in today's world. They are a slender and wobbly reed, and at all odds little inclined to collaborate'.[80] Wagar's view is based on two points. First, that not all 'anti-systemic' movements can be seen as being intrinsically opposed to the capitalist world-system, and second, that many such movements are co-opted, as some of their aims are adopted by the capitalist system in an attempt to silence their dissenting voices. This is a recurring problem of resistance, past and present. Other authors are more specific in their critique of the emerging resistance movements to globalisation. For example, Sklair[81] and Walker[82] both accept that local challenges to globalisation can be successful, however both seem to reject the view that social movements, such as the labour movement, can mount an effective global level challenge to economic globalisation. Walker's opposition to the view that social movements can resist globalisation on a global level is based upon his belief that social movements are restricted to the 'inner realms of society, in particular civil society.'[83] Sklair's argument is based upon his belief that social movements will never reach the high levels of international cooperation and coordination which would be needed to resist globalisation on anything other than a local or national level.

What then is the future for political resistance against neo-liberal globalisation? One thing is clear: the close relationship between the state and economic restructuring means that resistance to globalisation will continue to come predominantly from within national civil society and national social movements, including organised labour. The future success of resistance movements to neo-liberal globalisation may be brought a step closer if resistance organisations themselves highlight the close relationship between the state and globalisation. Resistance groups should act to break down the myth, which is often perpetrated by governments, that they are helpless in the face of globalisation, and refuse to accept that their own hands are tied by the inevitable on-rush of global economic forces. Resistance movements need to continue to stress the links between global restructuring and other issues of public discontent: for example, widespread unemployment, environmental degradation, rising malnutrition and loss or decline of health care services. A better understanding of the linkages between the state and economic restructuring and the breakdown in the social fabric needs to be popularised, as do social forces' own essential role in resisting neo-liberal globalisation. Both of these tasks are necessary in promoting widespread popular opposition to neo-liberal globalisation, which must occur if it is to be successfully challenged.

These tasks mean also that resistance movements need to break out of the local. The local is the site of much resistance, and is a source of great strength to resistance movements, as feelings of common culture, community, and of a shared sense of suffering can be used to mobilise action. However, primary or exclusive emphasis on the local can also lead groups to become colloquial and blinkered to other acts of resistance around the world or even in their own region, leaving them exposed to defeat or even destruction due to insufficient social alliances. The tragedy of the Ogoni people in Nigeria recently is an example of what can happen to resistance movements that are too preoccupied with the local scale of resistance, yet operate against global economic forces allied to the national state.

To be successful, resistance to neo-liberal globalisation must be conducted in a coordinated manner on a local, national, regional, and global level. Global restructuring is occurring on all levels, therefore resistance movements cannot defeat it by concentrating on one level alone; capital can always side-step such opposition. History has shown that once resistance becomes marginalised it will fail. If a firm is global, then labour unions must be global as well. The 'global strike' by workers in the same company and industry is a tactic yet to be deployed, and requires the preparatory stage of bringing such workers and their unions together to forge unifying strategies of resistance. Whereas the Ogoni were all too easily isolated and destroyed, due perhaps in large part to insufficient alliances in their own region, the Zapatistas survived and continue to organise, due perhaps to their better preparatory work in their own region, that is, alliances with other popular organisations (the EZLN's National Democratic Convention attracted some 6,000 delegates from all over Mexico), and to the fact that they are an armed popular movement capable of self-defense.

With regards to the future development of global resistance by 'new' social movements, the women's health movement may provide us with a useful example. The women's health movement is indeed a global social movement, which has successfully campaigned to make reproductive rights the centrepiece for international population policy in the 21st century. Its success is due to the fact that it spanned both the developed and developing world and mobilised on a local, national, regional and global level to get its views adopted. It has effectively used the technique of lobbying international organisations, such as the UN, and by direct pressure and participation succeeded in changing the political agenda of such organisations. Resistance movements which are opposed to globalisation can follow this lead and become organised and develop alliances and linkages with other resistance movements on the local, national, regional and global level.

The emergence of global popular forums of resistance to globalisation, such as the *International Forum on Globalization*, are a part of this process, as such forums provide a meeting place for resistance movements and are an important conduit for co-ordination.[84] There are likewise a number of organisations whose express purpose is to coordinate resistance on a wider scale than national movements, such as the Maquila Solidarity Network (MSN), Women Working Worldwide (WWW), the International Centre for Trade Union Rights (ICTUR), and the International Restructuring Education Network Europe (IRENE)[85] to name only a few.

To be successful, resistance groups must offer a viable alternative to neo-liberal globalisation. Democracy, sustainable development, welfare and social justice are values that may form part of the necessary basis for such an alternative political economy. It is important to recognise the wider influence resistance movements may be capable of achieving. Studies have shown that resistance movements to globalisation can also be an important source of democracy[86] and also a source of new knowledge and epistemology.[87]

CONCLUSION

In conclusion, it is our view that resistance from within civil society and from the base of organised labour and new social movements is an essential part of the challenge to neo-liberal economic globalisation. At present civil society continues to be primarily a national phenomenon, but an incipient popular 'international civil society' does exist and may further develop into a potent new force for change. To be successful, resistance must continue to have its roots in civil society, but must re-invent its relationship to the state and the political process by transcending traditional national boundaries through networks of solidarity, building regional and international resistance movements based on coordination and linkage.

In the past social movements and organised labour have been more confined to the local and national level, sustained as they are by the popular myths of the nation-state (i.e. that each state creates its own national society), than capital, which has always been more capable of escaping the 'national' to promote its interests. Given the extension of capital's 'flight' from the nation-state, it is more necessary and urgent today perhaps than ever before that resistance movements attempt to achieve a similar scale of operation. Paradoxically, the current discourse on globalisation and the 'end of the nation-state', that is, the new ideology of expanding capital, actually provides an opportunity for popular resistance movements to escape the myths of the nation-state. Resistance must move from the merely reactive (to capital accumulation processes) to being proactive and transformative. Therefore, resistance to neo-liberal globalisation must proceed in a coordinated and forward-looking manner; piecemeal and reactive resistance to global forces of capital accumulation is more easily coopted, marginalised and defeated.

If we take the neo-Gramscian project of extending Gramsci's ideas on hegemony from the domestic level to the international level seriously, then we have to realise that the consensual institutions of which Gramsci spoke are not only among international elites, but largely still embedded within what is usually seen as the 'national society' or civil society. It is the task of movements grounded in civil society to change the direction of state practice in a direction that counters the neo-liberal globalisation trend. It is from the unifying of these movements and their strategies that a counter-hegemonic potential must be found. Therefore, we must develop concrete strategies and concrete forms of organisation to resist the damage inflicted on society and the environment by the expansion of capital accumulation and the market. It is time to begin to 'overturn neo-liberal globalisation', and by doing so create a new world order.

NOTES

1. The editors would like to thank Carfax for permission to reprint this article originally published in *New Political Economy* (Vol.2, No.1, March 1997), pp. 179–95.
2. The authors are members of the Newcastle Research Working Group on Globalisation, of the Department of Politics, University of Newcastle upon Tyne. We would also like to acknowledge the participation and contribution of Carlo Berton and David Miller, and the helpful comments provided by Randall Germain.
3. For further discussion of the historical processes that contribute to present concepts and perceptions of 'globalisation' see Craig N. Murphy, *International Organization and Industrial Change: Global Governance Since 1850* (Cambridge: Polity Press, 1994).
4. Larry Elliott, 'Putting trade in its place', *The Guardian* (27 May, 1996), p. 14.
5. *Ibid.*
6. J.K. Galbraith, *The World Economy Since the Wars* (London: Sinclair Stevenson, 1994), pp. 247–8.
7. Ralf Dahrendorf, 'Preserving Prosperity', *New Statesman and Society* (15–29th December, 1995), pp. 35–40.
8. See, for example, Stephen Gill, 'Globalisation, Market Civilisation and Disciplinary Neoliberalism,' *Millennium: Journal of International Studies* (Vol. 24, No. 3, 1995), pp. 399–23.
9. We are indebted to Hugo Radice for bringing this point to our attention.
10. See Robert W. Cox, 'A Perspective on Globalization,' in James Mittleman (ed.), *Globalization: Critical Reflections* (Boulder, CO: Lynne Rienner, 1996), pp. 21–32.
11. Mark Beatty, IPE Discussion, Internet: 'G'obalization: What does it mean and is it good or bad?' 8 Oct. 1996; ipe@csf.colorado.edu; Beatty.4@osu.edu.
12. See, for example, Michel Aglietta, *A Theory of Capitalist Regulation* (London: New Left Books, 1979); Bob Jessop, 'Fordism and Post-Fordism: Critique and Reformulation' in Michael Storper and Allen J. Scott (eds.), *Pathways to Industrialization and Regional Development* (London: Routledge, 1992).
13. David Harvey, *The Condition of Postmodernity* (Oxford: Basil Blackwell, 1989). Jean-François Lyotard, *The Postmodern Condition* (Manchester: Manchester University Press, 1986).
14. Manuel Castells, *The Information City: Information Technology, Economic Restructuring, and the Urban Regional Process* (Oxford: Basil Blackwell, 1989).

15. Manuel Castells and Peter Hall, *Technopoles of the World: The Making of 21st Century Industrial Complexes* (London: Routledge, 1994), p. 3.
16. Jessop, *op. cit.*
17. Michael Piore and Charles Sabel, *The Second Industrial Divide: Possibilities for Prosperity* (New York, NY: Basic Books, 1984).
18. R.J. Barry Jones, *Globalisation and Interdependence in the International Political Economy: Rhetoric and Reality* (London: Pinter, 1995).
19. Kenichi Ohmae, *The Borderless World: Power and Strategy in the Interlinked Economy* (London: Collins, 1990).
20. Richard B. McKenzie and Dwight R. Lee, *Quicksilver Capital: How the Rapid Movement of Wealth has Changed the World* (New York, NY: Free Press, 1991).
21. Susan Strange, *States and Markets: An Introduction to International Political Economy* (London: Pinter, 1988).
22. Stephen Gill and David Law, *The Global Political Economy* (Brighton: Harvester Wheatsheaf, 1988).
23. Roland Robertson, *Globalization: Social Theory and Global Culture* (London: Sage, 1992).
24. Michael Porter, *The Competitive Advantage Of Nations* (Basingstoke: Macmillan Press, 1992).
25. Robert Reich, *The Work Of Nations: Preparing Ourselves for 21st Century Capitalism* (London: Simon and Schuster 1992). Philip G. Cerny, *The Changing Architecture of Politics: Structure, Agency and the Future of the State* (London: Sage, 1990).
26. Anna Pollert, *Farewell to Flexibility* (Oxford: Basil Blackwell, 1991).
27. *The Economist*, 'The Myth of the Powerless State' 7 October 1995, pp. 15–16.
28. Piore and Sabel, *op. cit.*
29. 'La Technologie et l'economie, les relations determinantes' OCDE, Paris, 1992, 'sur les facteurs techniques de la globalisation'.
30. Jones, *op. cit.*
31. Razeen Sally, 'Multinational Enterprises, Political Economy and Institutional Theory: Domestic Embeddedness in the Context of Internationalisation,' *Review of International Political Economy* (Vol. 1, No. 1, 1994), pp. 161–92.
32. Geoffrey Hodgson, *The Economics of Institutions* (Oxford: Edward Elgar, 1994). Neil. J. Smelser and Richard Swedberg (eds.),*Handbook of Economic Sociology* (Princeton, MA: Princeton University Press, 1994).
33. John Zysman, 'The Myth of "Global" Economy: Enduring National Foundations and Emerging Regional Realities,' *New Political Economy* (Vol. 1, No. 2, 1996), pp. 157–184. See also Gosta Esping-Andersen (ed.),*Welfare States in Transition: National Adaptations in Global Economies* (London: Sage, 1996).

34. See Mittleman, *op. cit.;* Winifred Ruigrok and Rob van Tulder, *The Logic of International Restructuring: The Management of Dependencies in Rival Industrial Complexes* (London: Routledge, 1995); Jones, *op. cit.*
35. *Ibid.*
36. Will Hutton, *The State We're In* (London: Jonathan Cape, 1995).
37. Roland Robertson, 'Mapping the Global Condition: Globalization as the Central Concept,' *Theory, Culture and Society* (Vol. 7, No. 1, 1990), pp. 15–30. For a discussion and critique of the 'transnational model' in the study of the globalisation of finance, see Benjamin Cohen, 'Phoenix Risen: The Resurrection of Global Finance,' *World Politics* (Vol. 48, 1996), pp. 268–96.
38. Susan Strange, 'Finance, Information and Power,' *Review of International Studies* (Vol. 16, No. 2, 1990), pp. 259–74; Philip Cerny, 'The Infrastructure of the Infrastructure? Towards "Embedded Financial Orthodoxy" in the International Political Economy', in Ronan Palan and Barry Gills (eds.), *Transcending the State-Global Divide: A Neostructuralist Agenda in International Relations* (Boulder, CO: Lynne Rienner, 1994), pp. 223–50.
39. Robert Reich, *The Work of Nations: A Blueprint For the Future* (London: Simon and Schuster, 1991).
40. Robertson, *op. cit.*
41. Susan Strange, 'The Defective State', *Daedalus* (Vol. 124, No. 2, 1995), pp. 55–74.
42. Cerny, *op. cit.*
43. Reich, *op. cit.*
44. Strange, *op. cit.*, pp. 55–7.
45. Zysman, *op. cit.*, pp. 157–84.
46. Andrew Cox, 'The State, Finance Industry Relationship in Comparative Perspective', in Andrew Cox (ed.), *The State, Finance and Industry* (Brighton: Harvester Wheatsheaf, 1986), pp. 1–59.
47. Zysman, *op. cit.*
48. Paul Hirst and Grahame Thompson, *Globalization in Question* (Cambridge: Polity Press, 1996); Jones, *op. cit.*
49. Ruigrok and Van Tulder, *op. cit.*
50. Geoffrey Garret and Peter Lange, 'Political Responses to Interdependence: What's "Left" for the Left?', *International Organization* (Vol. 45, No. 4, 1991), pp. 539–64.
51. Alain Lipietz, *Mirages and Miracles: The Crisis in Global Fordism* (London: Verso, 1987).
52. Zysman, *op. cit.*, p. 164.
53. Hirst and Thompson, *op. cit.*
54. Leo Panitch, 'Globalization and the State', *Socialist Register: Between Globalism and Nationalism* (Manchester: Merlin Press, 1994), pp. 60–93.
55. Karl Polanyi, *The Great Transformation: The Political and Economic Origins of Our Time* (Boston, MA: Beacon Press, 1944).

56. Saskia Sassen, 'The State and the Global City: Notes Towards a Conception of Place Centered Governance', *Competition and Change: The Journal of Global Business and Political Economy* (Vol. 1, No. 1, 1995), pp. 31–50.
57. Sam Pooley, 'The State Rules, OK? The Continuing Political Economy of Nation-States', *Capital and Class* (Vol. 43, No. 1, 1991), pp. 65–82.
58. Cerny, *op. cit.*
59. Andre Drainville, 'International Political Economy in an Age of Critique of Open Marxism', *Review of International Political Economy* (Vol. 1, No. 1, 1994); Zysman, *op. cit.*
60. Much of this academic interest has come from scholars within the disciplines of International Relations and International Political Economy. For example Robert W. Cox, *Production, Power and World Order: Social Forces in the Making of History* (Washington, D.C.: Columbia University Press, 1987); Martin Shaw, 'Civil Society and Global Politics: Beyond a Social Movement Approach', *Millennium: Journal of International Studies* (Vol. 23, No. 3, 1994), pp. 647–76; and Ronnie D. Lipshutz, 'Reconstructing World Politics: The Emergence of Global Civil Society', *Millennium: Journal of International Studies* (Vol. 21, No. 3, 1992), pp. 389–420.
61. Roger Simon, *Gramsci's Political Thought – An Introduction* (London: Lawrence and Wishart, 1991), p. 27.
62. Ragnu Krishnan, 'December 1995: "The First Revolt Against Globalization"', *Monthly Review* (Vol. 48, No. 1, 1996), pp. 1–22.
63. Jamie Morales, 'Indigenous Movements and the Fight for Land in Mexico' 1996, *INTemas* (Vol. 3, No. 7, 1996), pp. 15–17.
64. Patrick Bond and Mzwanele Mayekiso, 'Toward the Integration of Urban Social Movements at the World Scale', *Journal of World-Systems Research* (Vol. 2, No. 2, 1996), pp. 1–10. SANCO (South African Civic Organisation) is a national umbrella association of the various township civic associations which exist within South Africa.
65. Richard Dodgson, 'Global Social Movements', paper presented at the colloquium on *Political and Economic Restructuring in a Post-Modern Era*, Department of Politics, University of Newcastle upon Tyne, 27 May 1996.
66. June Nash and Christine Kovic, 'The Reconstitution of Hegemony: The Free Trade Act and the Transformation of Rural Mexico', in Mittleman, *op. cit.*, pp. 165–85.
67. Kenna Owoh, 'Fragmenting Health Care: The World Bank Prescription for Africa', *Alternatives* (Vol. 21, No. 2, 1996), pp. 211–37. Also see the African NGO Declaration for UNCTAD IX, posted to wsn@csf.colorado.edu and ipe@csf.colorado.edu, 4 May 1996 by Patrick Bond pbond@wn.apc.org, National Institute for Economic Policy, Johannesburg.
68. See the special section on international labour standards in *New Political Economy* (Vol. 1 No. 2, 1996).

69. For example, the Zapatista Front for National Liberation (EZLN) National Democratic Convention was attended by 6,000 delegates from all over Mexico. Moreover, via a number of international support groups the EZLN wins international support and recognition for its actions.
70. Bond and Mayekiso, *op. cit.*, p. 9.
71. Christopher Chase-Dunn, World Systems Network, Internet, wsn@csf.colorado.edu, chriscd@jhu.edu, 'Buchanan's right on the New World Order', 7 May 1996.
72. Robert W. Cox, 'Global Perestroika', in Robert W. Cox and Timothy J. Sinclair, (eds.) *Approaches to World Order* (Cambridge: Cambridge University Press 1996), p. 310.
73. John Cavanagh (ed.), *Beyond Bretton-Woods: alternatives to the global economic order* (London: Pluto Press, 1994), p. xviii.
74. Richard Falk, 'An Inquiry into the Political Economy of World Order', *New Political Economy* (Vol. 1, No. 1, 1996), pp. 13–27.
75. Bond and Mayekiso, *op. cit.*, p. 9.
76. See Andre Drainville, 'Left Internationalism and the Politics of Resistance in the New World Order', in David A. Smith and Jozsef Borocz (eds.), *A New World Order? Global Transformations in the Late Twentieth Century* (London: Praeger, 1995).
77. Chase-Dunn, *op. cit.*
78. Angela Hale, 'The deregulated global economy: women workers and strategies of resistance', *Gender and Development* (Vol. 4, No. 3, 1996), pp. 8–15.
79. Drainville, *op. cit.*, pp. 226–30.
80. W. Warren Wagar, 'Towards a Praxis of World Integration', *Journal of World-Systems Research* (Vol. 2, No. 2, 1996), p. 8.
81. Leslie Sklair, 'Social Movements and Global Capitalism', *Sociology* (Vol. 29, No. 3, 1995), pp. 495–512.
82. R.B.J. Walker, 'Social Movements/World Politics', *Millennium: Journal of International Studies* (Vol. 23, No. 3, 1994), pp. 669–700.
83. *Ibid.*, p. 672.
84. J. Cavanagh, (ed.), 'South-North: Citizen Strategies to Transform a Divided World', *IFG White Paper No. 1* (San Francisco, CA: The International Forum on Globalization, 1996).
85. International Restructuring Education Network Europe (IRENE) works to develop strategies and campaigns to protect worker's rights worldwide and to develop a code of conduct for TNCs. The Maquila Solidarity Network (MSN) promotes solidarity between labour and social movement groups between Canada and Mexico and Central America, and works to develop organising strategies that connect community and workplace issues, address health and environmental issues and the problems of women workers in the maquiladora. International Centre for Trade Union Rights (ICTUR) works to

defend and extend the rights of trade unions and workers worldwide. Women Working World Wide (WWW) supports women workers through international networking and public education and studies the impact of liberalisation on women workers in many countries.

86. Glenn Adler, 'Global Restructuring and Labour: The Case of the South African Trade Union Movement', in Mittleman, *op. cit.*, pp. 117–43.

87. Marianne H. Marchand, 'Latin American Voices of Resistance: Women's Movements and Development Debates', in Stephen J. Rostow, Naeem Inayatullah and Mark Rupert (eds.), *The Global Economy as Political Space* (Boulder, CO: Lynne Rienner, 1994), pp. 127–44.

6. Theorising Politics in 'No Man's Land': Feminist Theory and the Fourth Debate
Bice Maiguashca[1]

Those unhappy with the predominance of positivism in international relations and seeking to find new directions within this generally conservative discipline are facing hard times. The present champions of alternative theorising are in disarray, with three main groups – critical theorists, postmodernists and feminists – arguing with each other over who should lead the forces of post-positivism in the discipline. This small contingent of discontented international relations theorists, while agreeing that their positivist colleagues are wrong, are trapped in a virulent debate among themselves over the fundamental questions of political subjectivity, the status of knowledge, and the possibility of emancipatory politics. As critical theorists and postmodernists, the two loudest voices in this debate, increasingly move in opposite directions, feminists are left to talk among themselves. Yet, if one examines closely the work of feminist theorists, both outside and inside international relations, one discovers that it is precisely their work that has the most potential to steer the post-positivist debate beyond its current impasse and, in so doing, to offer a new conceptualisation of politics in international relations.

This potential contribution stems from the fact that feminist theorists, more than their post-positivist colleagues, have grounded their theoretical work in the particular political practices of specific social movements and have sought to respond to the concrete needs of an identified addressee, that is, women. Accordingly, feminists have approached the post-positivist debate from a far more pragmatic perspective and have tended to draw on both critical theory and postmodernism according to need. This practical orientation, in turn, has led some feminists to argue for a more *integrative* approach which combines the concerns and insights of both these perspectives and which is based on empirical research. It is in the efforts of these feminist theorists that I see the greatest promise for a feminist contribution to a post-positivist international relations theory.

This chapter is divided into four parts. In the first I will characterise the debate between critical theorists and postmodernists, or what I shall call the Fourth Debate, for it is within this context that recent discussions about the theorisation of politics and the possibility of a post-positivist agenda for international relations have taken place. In the second, I will locate feminist theory within this contested theoretical field and show how feminists, while replicating this debate in a number of ways, do so from a much more pragmatic vantage point. In the third I will explore the work of Nancy Fraser, a feminist philosopher, whose work learns from both sides of the post-positivist divide and goes beyond that of international relations feminists in the formulation of a theory of politics. Arguing that the critical theory versus postmodern debate is based on a false antithesis, she proposes that the way forward is through the development of a social theory that focusses on social movements and is based on empirical research. In the final part I will outline an alternative agenda for the study of politics in international relations based on the insights offered by feminist theory in general, and Fraser in particular.

THE FOURTH DEBATE: THE CRITICAL THEORY VERSUS POSTMODERNISM STALEMATE

Since the 1980s international relations theorists have been engaged in what has come to be known as the Third Debate. Originally this debate was seen as a confrontation between different ontological perspectives (realists, pluralists and structuralists).[2] Later, Mark Hoffman and Nicholas Rengger suggested that it could also be understood as a clash between different epistemological perspectives. This time the fundamental dividing line among the contending parties was drawn between 'positivists', represented by realists, pluralists and some structuralists, and 'post-positivists' represented by critical theorists, postmodernists and feminists, although they are curiously silent about this last group.[3] What I shall call the Fourth Debate refers to the ongoing discussion *within* this post-positivist camp. As stated above, this debate has been largely dominated by critical theorists and postmodernists with feminists, on the whole, being relegated to the margins.[4] For the purposes of this chapter, therefore, the Fourth Debate shall refer exclusively to the ongoing conflict between these two groups.

Critical theory and postmodernism have a great deal in common, a fact often forgotten in the vitriolic exchanges that characterise the discussion between these two camps. After all, both critical theorists and postmodernists are engaged in an effort to critique prevailing social and political practices and to reveal the power relations that underwrite them. To this extent, both are interested in understanding the nature of domination and the possibilities for less oppressive social relations. Moreover, both are concerned with the question of social change and the role of social movements within this context. More specifically, they are politically motivated and choose to focus their theoretical attention on the experiences of those who have been marginalised by the dominant society. Finally, both point to the relationship between knowledge and power and the importance of recognising the historicity of knowledge and, thereby, the partial nature of all truth claims.

These commonalities notwithstanding, critical theorists and postmodernists offer very different narratives of oppression and emancipation. According to the former,[5] we live in a world of *material* social structures which produce unequal social relations in which some gain at the expense of others. The central structure which dominates our lives today and which is the primary cause of these oppressive relations is the global capitalist economy. It is this socio-political, economic order which transforms social relations into impersonal, money-oriented exchanges and which reproduces a variety of oppressive relationships, such as those based on sexism, racism and ethnocentrism. Some critical theorists, those of a Habermasian persuasion,[6] go beyond this point and focus on how the capitalist logic has penetrated our cultural-ethical sphere (or as Habermas calls it, the 'life world'). In their view, this penetration is responsible for the dominance of 'instrumental reason' over our more ethical, emancipatory inclinations, which are sustained instead by 'communicative reason'. The liberation of humanity, therefore, depends on breaking the grip that the capitalist system and, even more importantly, its instrumentalist logic have on our lives both as productive and communicative beings.

The impulse for emancipatory change, critical theorists argue, springs from our 'critical reason' which allows us, as knowing, self-conscious human beings, to envision and act upon an alternative political agenda. One of the central goals of the critical theory project, therefore, is the recovery of our capacity for 'critical reason', which in our age, as mentioned above, has become subservient to 'instrumental reason'. In this connection, the task of the theorist is to critique and de-legitimise prevailing ideological beliefs which serve to obscure the oppressive reality of capitalist society and in this way maintain the status quo. Without this ground-clearing exercise, it is impossible for those acting for social change to know where and when to act and, indeed, to envision a new, more emancipated social world.

Postmodernists have challenged this narrative of critical theorists and have focussed their efforts on demystifying and destabilising this project of reason and emancipation.[7] They point out that it is reason itself, not capitalism, which is responsible for much of what is dark and oppressive about our world today. For the postmodernists, then, the problem begins with Enlightenment philosophy and modern rationality which, they argue, seeks to rationalise, categorise and dominate all within its grasp, rather than with a particular social arrangement or political practice. Accordingly, it is only with the rejection of Enlightenment philosophy that a meaningful social critique is possible.

Each of these two different visions of oppression and emancipation rests on three metatheoretical assumptions and it is mainly around these three axes that the Fourth Debate has developed and revolved. The first concerns ontology and relates to their contrasting views of the human subject. Critical theorists assume a fundamental divide between mind and matter or between the world of (inter)subjectivity and the world of things. Human subjects for critical theorists are by their nature centred, unified, *social* beings endowed with specific needs and interests which are anchored on non-contingent forces. They are self-conscious, self-reflexive beings, capable of self-transformation. Thus, critical theorists begin their analysis of the social world with clear ontological assumptions, which are both *universal* and *transcendental* in nature.

Postmodernists, on the other hand, portray the human subject as fragmented and decentred and as constituted by a variety of social and psychological forces beyond her understanding or control. Furthermore, for postmodernists an individual *becomes* a 'subject', that is, gains an identity through her actions. Identity is generated through repeated practices, or to use their own words, it is 'performatively' produced. Consequently, ontology is left aside and is replaced by a concern with mapping the connections between contingency and a plethora of temporary, changing subjectivities. In sum, while the subject of critical theory has an ontological foundation, the protagonist of postmodernism does not.

The second axis around which the debate between critical theory and postmodernism revolves is epistemology. For the critical theorist, critique and that which enables it, universal critical reason, is a prerequisite for emancipatory political practices. Thus, although knowledge is generally understood as historically and socially produced and, therefore, open to revision in the future, critical reason must be granted some ahistorical or privileged status so that it can serve as the barometer of 'correct' truths and 'real' emancipatory change. Without this status, universal critical reason cannot be a medium through which the project of self-liberation can be articulated or the means by which to determine the criteria for such a project.

Postmodernists, on the other hand, refuse to privilege any one form of knowledge or reason over others: they believe that there is no universal, ahistorical criteria for assessing the validity of truth claims. In other words, there is no foundation on which to secure critical reason or emancipatory knowledge. In fact postmodernists go further and argue that the 'will to truth' cannot be separated from the 'will to power' and that the formation of knowledge discourses is the result of exclusionary practices. Thus, for postmodernists, supposedly emancipatory discourses, such as the 'myth of reasoning, sovereign man', can be just as oppressive as those which are explicitly exclusionary. In this way, they transform epistemology into a political exercise; what we know and how we know becomes a matter of power rather than objective reason, critical or otherwise.

The third axis of conflict between critical theorists and postmodernists has to do with power and transformative political practice. Critical theorists conceive of power as stemming from one central source (e.g. production relations) and of politics as the collective effort of an oppressed group to overturn the system as a whole for the benefit of all members of society. This is possible because human beings are guided by emancipatory interests, such as those of solidarity, redistribution and self-determination, and are capable of acting upon them. Critical theorists, therefore, understand politics as an activity which belongs both to the realm of power as well as to that of universal ethical principles.

By contrast, postmodernists offer a more dispersed and ubiquitous view of power. For them one important locus of power is language, which is understood not as a neutral medium, but rather as a power-ridden playing field which conditions what can be said and, more importantly, what cannot. Language also plays an important role in both the formation of social identities and in their politicisation. Thus, unlike critical theorists, who begin their analysis with a clear and unambiguous idea of the centre of power, postmodernists focus on what they call 'discursive relations', that is, the complex web of power-laden discourses that permeate, at all levels, a particular social order. As for politics, it is a localised and contingent affair and involves the reappropriation and rearticulation of specific 'discursive regimes' rather than the overthrow of societal structures *in toto*. Thus politics, with no ontological or epistemological foundations, is relegated by the postmodernists to the realm of contingency and revolves around the enactment of alternative identities not legitimised by the status quo.

In sum, despite their attention to the historicity of social and political life, critical theorists lean on the twin pillars of universalism and transcendentalism when they conceptualise the 'sovereign' subject, critical reason and the meaning of emancipatory politics. That is to say they adhere to what is called a foundationalist perspective. Postmodernists, on the other hand, swear off these two foundational touchstones, preferring to let historicity take its course. As Walker makes clear, '[u]niversalism, to put it bluntly and heretically, can be understood as the problem, not the solution.'[8] As can be seen, the two main contenders of the Fourth Debate have reached an impasse. It is the claim of this chapter, however, that this debate can be revitalised if one listens to recent feminist voices inside and outside the discipline who are suggesting that there is a way beyond the current stalemate.

FEMINIST THEORY AND THE FOURTH DEBATE

While there is a growing body of feminist literature within international relations, as of yet it is still underdeveloped, especially in terms of metatheory. Indeed, those international relations feminist theorists who do address metatheoretical questions, such as Christine Sylvester, draw very heavily on their feminist colleagues in the disciplines of political theory and philosophy. For this reason I too cast my net beyond international relations in order to elucidate, in more detail, the feminist version of the Fourth Debate.[9] The main protagonists in the feminist debate have been critical theory feminists pitted against postmodern feminists. Within the first camp, there are those who draw their ideas from the Frankfurt School and Habermas in particular, the 'Habermasian' critical theorists, and those who borrow mainly from Marx, the 'standpoint' feminists. Although I address them separately, I group them together under the label of 'critical theorists' since both subscribe to a foundationalist perspective. As for the postmodernist feminists, I also group them together given their common anti-foundationalist stance.

Despite their differences, what shall become clear is that feminist scholars of all theoretical persuasions have remained united around their central concern with theorising and empowering the women's movement. It is this common goal that has enabled them to appreciate each other's insights and encouraged some of them to seek a more integrative approach to conceptualising politics.

Reflecting the Fourth Debate, the feminist debate has centred on three main themes: the notion of the human subject, the status of truth and the appropriate theoretical grounding for feminist politics.

Notions of the Human Subject

Not all feminist critical theorists share the same feminist ontology.[10] Those more inclined to a standpoint perspective, such as Nancy Hartsock, view women's subjectivity as defined predominantly by their position in the sexual division of labour. Given women's role in the home and, specifically in child rearing, Hartsock argues that, ontologically speaking, women can be conceptualised as 'relational beings' who have a 'sense of a variety of connectednesses [sic] and continuities both with other persons and with the natural world.'[11] For those more influenced by Habermas, such as Seyla Benhabib, on the other hand, the subject is defined in terms of her capacity to engage in communicative practices which, in turn, provide the context in which her identity as a moral being is constructed.[12] Despite these differences, what both Hartsock and Benhabib do share is a belief that without a concept of 'woman', as a subject capable of autonomy, self-determination, self-reflexivity and consciously-guided action, it is impossible to conceptualise women's agency and, therefore, the women's movement.

Postmodern feminists acknowledge the important role that the category of 'women' plays in feminist politics, but they also insist on the need to destabilise it in an effort to free it from its restrictive, foundationalist moorings. Judith Butler, for example, argues that the category 'women', as used by critical theorists, is exclusionary in nature. As a set of values and feminine traits, it depicts a human subject that is not representative of all women and therefore marginalises those who do not fit into this imposed model. For Butler, gender identity cannot be conceived as pregiven, self-evident or self-transparent; rather, it must be understood as an *effect* of psychological, political and cultural processes which creates only the *appearance* of a unified identity.

The loss of the ontological category of 'women' is not a problem for Butler in terms of feminist politics because it is replaced by the notion of 'subject-positions' which allows for the study of a network of coalitions that do not presuppose constitutive subjects. As Butler concludes, 'it seems crucial to resist the myth of interior origins, understood either as naturalized or culturally fixed. Only then, gender coherence might be understood as the regulatory fiction that it is – rather than the common point of our liberation.'[13]

The Status of Truth

Turning to epistemology, although both feminist critical theorists and postmodern feminists agree that there is an inherent relation between knowledge and power, they disagree about what this means for the status of truth and the possibility of emancipatory politics. In general terms, feminist critical theorists argue for the need to distinguish between oppressive and emancipatory forms of knowledge. Postmodernist feminists, on the other hand, insist on the inevitability of all knowledge claims turning into monolithic discourses of power.

More concretely, for feminists of a critical theory persuasion, the crucial question is: on what epistemological grounds can a feminist politics be built? Nancy Hartsock, the standpoint feminist, argues that the sexual division of labour offers one vantage point from which we can secure knowledge about women's lives and their politics because it is within this context that oppression and resistance take place. In this connection, Hartsock claims that the subjectivity of women '*qua* oppressed' offers a less distorted vision of the world, one which can lead to a reconstruction of social relations along more equitable lines. Apart from their material activities within the sexual division of labour, Hartsock pays attention to the importance of political activism as a crucial experience which shapes feminist consciousness.

Seyla Benhabib, a Habermasian feminist, also stresses the importance of political contestation as a source of knowledge:

> [s]uch understanding is a product of political activity. It cannot be performed either by the political theorist or by the moral agent *in vacuo*. For as Arendt well knew, the multiplicity of perspectives which constitute the political can only be revealed to those who are willing to engage in the foray of public contestation.[14]

Many postmodern feminists have taken issue with these epistemological claims. Jane Flax, for instance, argues that to assume women share epistemological commonalities because they all experience oppression is to deny the differences between women's experiences and their identities. If one takes these diversities into account, the political question becomes: which women's experiences does one privilege and, therefore, which women are to be selected as the representatives of all others?

A second problem identified by Flax has to do with the assumption that, as a subject, a woman is capable of knowing herself and interpreting her experience 'correctly'. Women, like men, are defined by internal discontinuities which prevent complete self-knowledge and a clear epistemological vision. Thus, she rejects the idea that women can make universal, correct knowledge claims about the world, even when talking about their own experience of it.[15]

A third problem for postmodernists is whether women's truth claims are necessarily better than those offered by others; is it not possible, they suggest, that the experience of oppression takes its toll and distorts their vision of truth? With no universal criteria to assess the truth or legitimacy of a knowledge claim, what is the point in trying to hold on to any objective notion of truth? In light of these three problems, postmodern feminists maintain that politics cannot be grounded in a foundational epistemology.

Empowering Feminist Politics

Feminist politics is concerned with resistance to and liberation from gender oppression. In this connection, feminist critical theorists maintain that no coherent, unified resistance can be launched unless the *structural* sources of inequality are identified and understood. Thus, they call for an historical analysis of 'patriarchy' understood either in terms of the sexual division of labour or the androcentric nature of our cultural sphere.[16]

Besides conceptualising the structural nature of gender domination, feminist critical theorists also call for a normative vision which can guide women's struggles. Without some universal normative principles to light the way, they argue, feminist politics will remain anchored to the parochial interests of its diverse constituencies and will be unable to foster solidarity. Moreover, in the absence of normative criteria, the justification of feminist politics in any form becomes impossible in the first place.

Postmodern feminists, like their critical theory peers, are fundamentally concerned with the empowerment of feminist politics. The challenge from their point of view, however, is to free feminist politics from its structuralist and universalist moorings and to give full reign to the diversity, spontaneity and creativity of women's activism. For postmodernists there is no one central source of oppression to be overcome and, therefore, no one type of feminist politics that can secure freedom. Indeed for them the strength of the women's movement lies in its diverse and local nature and its capacity for contingent coalition alliances.

For Butler, for example, politics is 'performatively' enacted so that the normative presuppositions behind it can only be articulated in and through action. In this way all political claims are made in a context of struggle and since one cannot prejudge the direction and outcome of the struggle there is little point in trying to secure in advance the normative foundations for a feminist politics. Indeed, Butler writes 'that a fundamental mistake is made when we think we must sort out philosophically or epistemologically our "grounds" before we can take stock of the world politically or engage in its affairs actively with the aim of transformation.'[17] Any attempt to do this must be seen as a denial of the political at best and a covert effort to limit the political imagination of its participants, at worst. Rather than wish away the political, Butler urges feminists to accept and negotiate the matrix of power relations in which they live. In this context, normative principles cannot serve as foundations, but only as pragmatic guidelines.[18]

The Fourth Debate Reconsidered

As can be seen, feminist theorists have clearly replicated the Fourth Debate, but with a difference. While the terms of this debate have been dictated by the polemical deliberations of a small academic community of international relations scholars, the primary impulse behind the feminist debate has come from an effort to respond to the political practices of an evolving social movement. It is women's movements and the goal of empowering feminist politics which has served to give direction and contextualise the theoretical and metatheoretical pronouncements of feminist scholars. According to Lennon and Whitford, this political commitment and sense of accountability is intrinsic to feminist theory.[19] As I argue later, it is this dimension of feminist scholarship which has enabled it to avoid the sterility of the Fourth Debate and to provide an alternative vision of politics. More concretely, this grounding of feminist theory in political practice has required feminists to be less rigid and more self-reflexive about their theoretical positions. As a result, in recent years, there has been a softening of the boundaries in the feminist debate.

Some feminist theorists even argue that the conflict between critical theory and postmodernism is based on a false dichotomy. One such theorist, Sandra Harding, suggests that all feminists should adopt a 'principled ambivalence' to both these traditions of thought. Denying that the tensions between critical theory and postmodernism are a result of error on either side, she insists that 'they reflect different, sometimes conflicting, legitimate political and theoretical needs of women today.'[20] The challenge, then, according to Harding, is to understand these approaches as potentially complementary. As Harding states, '[a]t this moment in history, our feminisms need both Enlightenment and postmodernist agendas – but we don't need the same ones for the same purposes or in the same forms as do white, bourgeois, androcentric Westerners.'[21]

Following this new trend within feminist political theory, a number of international relations feminist scholars are also questioning the need to swear allegiance to one side or the other of the Fourth Debate.[22] Marysia Zalewski and Christine Sylvester, for instance, argue that there is no necessary choice between the foundational approach of critical theory and the non-foundational approach of postmodernism.[23] On the whole, however, their efforts to negotiate these two perspectives remain undeveloped both empirically and theoretically. It is only by examining the work of Nancy Fraser, a feminist philosopher, that one can begin to imagine an alternative route which draws on both perspectives and, thereby, establishes the basis for a distinct feminist contribution to the Fourth Debate. Once again, however, we must go outside of international relations to find what we are looking for.

BEYOND THE FOURTH DEBATE: THEORISING POLITICS IN 'NO MAN'S LAND'

Nancy Fraser's work relates to the Fourth Debate in at least two ways. First, it addresses the metatheoretical impasse between critical theory and postmodernism and suggests a way of going beyond it. In effect, although her work is in the field of feminist political theory, the issues she addresses are the same as in the Fourth Debate, thus making her approach relevant to international relations. Second, arguing that metatheoretical discussions are limiting unless invigorated by social theory and empirical research, she puts forward a substantive social theory of feminist politics that, on the one hand, can breathe new life into the Fourth Debate and, on the other, can serve as a model to post-positivist theorists seeking to offer an alternative conceptualisation of politics in international relations. I shall briefly touch upon her views on metatheory before examining at greater length her theory of politics.

In her two articles 'Pragmatism, Feminism and the Linguistic Turn' and 'False Antithesis',[24] Fraser assesses the metatheoretical strengths and weaknesses of both feminist critical theory and feminist postmodernism and states that, while each perspective offers useful insights, neither one alone is adequate. Furthermore, she claims that the debate between them has been constructed around a series of false antitheses. Rejecting the either/or choices presented by both Benhabib and Butler, such as identity/difference, contingency/structure, foundationalism/relativism, Fraser's position is that 'feminists do not have to choose between Critical theory and poststructuralism; instead, we might reconstruct each approach so as to reconcile it with the other.'[25]

In this vein, Fraser seeks to conceptualise the human subject as both centred and decentred, constitutive and constructed, real and imagined:

> Complex, shifting, discursively constructed social identities provide an alternative to reified, essentialist conceptions of gender identity, on the one hand, and to simple negations and dispersals of identity, on the other. They thus permit us to navigate safely between the twin shoals of essentialism and nominalism, between reifying women's social identities under stereotypes of femininity on the one hand and dissolving them into sheer nullity and oblivion on the other.[26]

For Fraser all claims about 'women' should recognise the existence of women who mobilise around their gender conceived as a universal and binding identity. But they should also regard this solidarity as vulnerable and subject to contingent forces. In this way, the crucial political question of how to deal with conflicts of interests among different groups of women is opened up for scrutiny rather than theorised away by either the critical theorists' tendency to smooth over difference or the postmodern habit of blindly celebrating it.

It is, however, at the level of social theory, rather than metatheory, that Fraser seeks to build a more detailed integrative approach. This preference is important. For her the current debate between the feminist critical theorists and feminist postmodernists is premised on a false antithesis and there is little to be gained from its continuance. Instead, what needs to be done, according to Fraser, is to put this debate on a new footing which is grounded in historical and empirical research and which aims at the articulation of substantive social theory.

Moving on to the particulars of her social theory, she calls for the articulation of a complex research project (a 'democratic-socialist-feminist-pragmatism') which can address both the critical theorist's concern with material and institutional structures of inequality as well as the postmodernist's interest in the role of language and discourse in the legitimation or devaluation of particular practices and identities. In addition, such a project would need to offer a conceptualisation of how the material and institutional sphere, on the one hand, and the discursive sphere, on the other, interrelate and mutually reinforce each other.[27]

Fraser develops her social theory in *Unruly Practices: Power, Discourse and Gender in Contemporary Social Theory* (1989) and *Justice Interruptus* (1997). Explaining her approach she writes:

> A critical approach must be bivalent... integrating the social and the cultural, the economic and the discursive. This means exposing the limitations of fashionable neostructuralist models of discourse analysis that dissociate "the symbolic order" from the political economy. It requires cultivating in their stead alternative models that connect the study of signification to institutions and social structures. Finally, it means connecting the theory of cultural justice with the theory of distributive justice.[28]

Inspired by Marx's definition of critical theory as 'the self-clarification of the struggles and wishes of the age', Fraser develops her social theory around the analysis of contemporary social movements and, in particular, the women's movement.[29] By choosing social movements as her main unit of analysis, Fraser makes both a methodological point and a normative one. In terms of the former, she argues that it is only through the historical and empirical study of specific social struggles that one can build a theory of politics. In terms of the latter, she claims that the possibility of emancipatory social change in the future depends on the commitment of intellectuals to the self-liberatory struggles of today.[30]

Fraser coins two concepts to understand 'the political'.[31] The first refers to those issues handled directly by the institutions and agencies of the state. She calls this *official-political* and contrasts it with the issues associated with the family and the economy. The second concept has to do with those issues which are 'contested across a range of discursive arenas and among a range of different publics.'[32] This she calls the *discursive-political*. These two forms of the political, she suggests, are connected to the extent that an issue is not usually subject to state intervention unless it is first debated among a range of publics. Accordingly, nothing is intrinsically political and any issue can potentially be politicised if it becomes subject to contestation and debate.[33]

Having said this, however, she reminds us that the politicisation of an issue is not so easily achieved. According to Fraser there are two institutions that depoliticise social discourses: the family and the economic system. The first personalises issues and relegates them to the private domain and the second turns them into impersonal market imperatives. Neutralised in this way, the uncontested discourses become part of the commonsense of a society, the unquestioned premises upon which people act.

Politicised issues, on the other hand, are those which have broken free from the economy or domestic realm and have leaked beyond their confines into uncharted, non-hegemonic terrain. This terrain Fraser calls *the social*. Borrowing from Arendt, Fraser defines this space as 'a switch point for the meeting of heterogeneous contestants associated with a wide range of different discourse publics', a 'no man's land' in which a plurality of interpretations about needs can compete with each other for dominance.[34] Given that for Fraser capitalist societies are constructed around a variety of discourses about needs that are hierarchically related, these discourses are sites of struggle in which groups with unequal discursive and non-discursive resources compete to establish their respective interpretations as hegemonic.

In this context, 'the politics of needs interpretation', as Fraser puts it, involves three distinct, but interrelated moments. In the first 'oppositional social movements' struggle to establish or deny the *political status* of a given need. This effort may be challenged by those social groups who, for a variety of reasons, aim to reprivatise and depoliticise those issues which oppositional movements have sought to contest. This effort at retrenchment can be taken on by organised interests, such as businesses and trade unions, or by political lobby groups. I would also add to Fraser's list the possibility of non-progressive social movements playing a part in this backlash.[35] The second moment in the politics of needs is the struggle over the interpretation of the need, that is, the battle over definition. Once again, private organised interests are pitted against oppositional social movements who must themselves come to a mutually agreeable understanding on the interpretation of the need. It is at this point that agencies associated with the state may become involved in the fray. Finally, the third moment in the politics of needs is the struggle over their implementation.[36] Now a new social actor – what Fraser calls 'expert publics' – occupies centre stage. Members of the professional classes,[37] these experts specialise in 'problem-solving' and seek to translate newly politicised needs into objects of state intervention and control. In this way, experts turn potentially radical discourses into manageable claims which can be dealt with administratively. This process, in turn, tends to depoliticise, decontextualise and recontextualise needs within the status quo.

In her latest work, Fraser argues that the main task for critical thinkers today is to distinguish analytically between two sets of social movements and thus between two types of political claims. The first set of claims concerns the demand for redistribution in a world of social inequality. This political claim, she argues, forms the basis of what she calls the 'politics of redistribution', with the trade union movement being a case in point. The second set of claims concerns the recognition of identity and difference and manifests itself through what she calls the 'politics of recognition'. One example of this kind of claim can be found in the gay and lesbian movement. According to Fraser, the 'politics of redistribution', the traditional concern of critical theorists, and the 'politics of recognition', the focus of postmodernists, cannot be understood separately since movements about redistributive justice are often waged by people within minority groups who have been disadvantaged because of their racial, ethnic, or gender identity. Conversely, movements concerned with identity issues, such as the women's movement or the gay movement, often become initially politicised around redistributive issues, such as equal pay or equal rights to employment. In other words, according to Fraser, movements mobilised around gender or race involve claims for *both* recognition and redistribution and, as a result, demand the transformation of both the political economy which perpetuates marginalisation and the system of cultural valuation which engenders stigmatisation. The central task for critical political thinkers today, therefore, is to understand how economic disadvantage and cultural disrespect go hand in hand and what political dilemmas arise when we try to combat both.[38]

In short, Fraser urges us to examine the way in which the boundaries between 'the political', 'the economic' and 'the domestic' are constructed and the way they shift in response to discursive conflicts which take place within the realm of 'the social'. Politics, in this light, is understood as involving social struggles over not only scarce resources, material inequalities or electoral positions, but also competing interpretations of symbolic and cultural needs. Thus, Fraser offers us a vision of politics and social change that incorporates the structures of the *official-political*, on the one hand, and the *discursive-political* sphere, on the other.[39] Furthermore, she asks us to pay close attention to the way in which power relations overlap (gender, ethnicity, class) and generate a multivalent politics which cannot be understood in terms of one logic alone, whether it be that of recognition or that of redistribution.

In conclusion, Fraser challenges the current state of polarisation between critical theory and postmodernism at the level of metatheory and characterises it as a false dichotomy. She goes on to suggest that in order to go beyond this impasse the best way forward is to develop a substantive social theory that is pragmatic in orientation and based on substantive empirical research. In addition, this social theory would have to be sensitive to both the material (structural) and cultural (discursive) dimensions of politics, and flexible enough to respond to the changing political practices of social movements.

FEMINIST THEORY AND ITS CONTRIBUTION TO A POST-POSITIVIST AGENDA IN INTERNATIONAL RELATIONS

> There is more to international relations than gender, and while feminism has opened up the discipline to neglected themes, it will not steer the discipline very far.[40]

> But feminists, like those attempting to draw on cultural traditions that have been eclipsed by the pretensions of the most powerful, are always in danger of relapsing into claims of privileged access, of reproducing the cultural arrogance they seek to undermine.[41]

The first quotation belongs to Andrew Linklater, a critical theorist, and the second to R.B.J. Walker, a postmodernist, both participants in the Fourth Debate. Whereas Linklater has beckoned feminists over to join the neutral, gender free ranks of critical theorists, Walker has admonished them to beware of the dangers of treading the middle ground where lie the pitfalls of foundationalism. Clearly, then, for the main contenders of the Fourth Debate feminist theorists must ally themselves with one side or the other. But must they? Is there nothing useful to be learned from theorising politics in the 'no man's land' between critical theory and postmodernism? More concretely, for those of us interested in advancing the Fourth Debate beyond its current impasse, how can the pragmatic bent of feminists contribute to the formulation of a more comprehensive theory of politics in international relations? In the last section I consider these questions by drawing from Fraser as well as international relations feminist thinkers.

Let us begin with the metatheoretical issues highlighted in the first two parts of this chapter. In terms of ontology, this pragmatic orientation allows post-positivists to think about the human subject in a different way because it militates against any metatheory that presupposes a universally applicable notion of the human subject, whether that involves the affirmation of a specific ontology (as favoured by critical theorists) or its absence (as favoured by postmodernists). In this connection, although critical theorists and postmodernists proclaim their respect for the historicity of social life, both tend to impose, in advance of empirical research, their own theoretical categories on changing social and political subjects. A pragmatic approach which is grounded in empirical analysis, on the other hand, would prevent this theoretical authoritarianism as it would require the theorist to adapt her categories to the variable nature of the human subject: its centred and decentred, constitutive and constituted dimensions.

A second point with regard to the significance of feminist pragmatism for a post-positivist ontology has to do with the importance that feminists give to social movements and social relations in general: for them these are the stuff politics is made of. Although many scholars in the Fourth Debate would agree with feminists on this count, few have actually developed an alternative ontology based on this insight. One notable exception to this general trend is Robert Cox who has made a conscious effort to conceptualise the role of social forces in international relations and who, as a result, has produced some of the most stimulating work in the discipline.[42]

In this connection, it would be interesting to explore the relevance of Fraser's concept of the 'social' for post-positivists in international relations. Can the concept of 'international public sphere' help us to conceptualise the arena in which social actors such as the state, social movements and non-governmental organisations operate and exert their influence?[43] A related task would be to map out the various, shifting connections between *movements* at the sub-national, national and supranational level. Furthermore, there is a need to explore the relationship between movements and states without presuppositions about the nature of this relationship. While a number of social movements have actually sought to take over state power and have done so successfully, others have engaged the state more indirectly, although no less politically. Why and how these different strategies came about and the effect that they have had on our understanding of the political has yet to be studied in our discipline. Finally, but most importantly, a more comprehensive theory of politics in international relations would need to conceptualise these movements as the efforts of marginalised peoples to resist oppression and to bring about the *political* transformation of the prevailing social order.

Turning now to the status of truth, the pragmatic approach of Fraser and other feminists has shown us that no one type of knowledge can be privileged or universalised *a priori*. If we accept this premise, the task for post-positivists in international relations is that of examining empirically the role that different kinds of knowledge claims have in generating political consciousness and political solidarity. For example, one important line of inquiry would be to identify the *specific* role of 'communicative rationality' in the construction of a particular politics. On the other hand, to claim that this particular form of knowledge is or should be at work in all forms of emancipatory politics in the world at large is to establish arbitrarily its superiority over other epistemologies and forms of emancipatory politics.[44]

Lastly, referring to the nature of power, feminists tell us that we must use the conceptualisations of both structural as well as discursive approaches. This means that, for the post-positivists, there are at least three crucial sites of power to be explored, all of which have structural and discursive dimensions. I am referring to the state and the agencies of the state, the economy and relations of production and the cultural sphere in which relations of gender, race and sexuality, among others, are constructed.[45] Although a significant amount of work has been done on the structural dimensions of power with regard to the state (e.g. military, treasury, law courts) and the economy (e.g. capital flows, banks, transnational corporations), much less has been written on the discursive elements of these two realms such as the androcentric nature of western law and of liberal economic theory. As for the cultural sphere, with the exception of feminist scholars, it has generally been ignored.

Before we leave metatheory, however, a word on Harding's notion of 'principled ambivalence' is needed. As has been seen throughout this chapter, a number of feminist scholars insist that the insights of both critical theory and postmodernism be seen as in some sense complementary rather than as mutually exclusive. The problem is how to act upon this premise without falling into mindless eclecticism. One of the central tasks of post-positivist international relations theorists, therefore, is to develop this notion of 'principled ambivalence' and, in so doing, set out the rules for responsible theoretical borrowing. To accomplish this, however, as Fraser has argued, it is imperative to move to the level of social theory, for it is only from this vantage point that one can justify the metatheoretical tools we choose.

Unfortunately, substantive social theory is not at the forefront of international relations theorising, post-positivist or otherwise. In fact, as stated earlier, the Fourth Debate has revolved mainly around metatheoretical issues. Thus despite some important initial contributions made to a substantive social theory of politics by international relations critical theorists such as Cox and Linklater, and international relations postmodernists such as Walker, at present the discipline of international relations remains largely devoid of any conceptual language that would enable us to identify and talk about social oppression, social resistance and social movements. Without such conceptual tools, I believe that it is simply impossible to speak about contemporary politics in our discipline.

Given this state of affairs, what are our alternatives? If we decide to follow through with the social theoretical work that does exist in the discipline we are faced with what Fraser has called a false dichotomy, that is, we must choose between critical theory and postmodernism. The former, on the one hand, offers us either a class-based notion of politics, which primarily focusses on social groupings arising out of production relations (Cox) or a notion of politics based on the moral learning of states (Linklater). The latter, on the other hand, presents us with a notion of the 'politics of movement' or the 'politics of difference', both of which focus on the contingent alliances of a variety of social movements. In the case of critical theory, the question of identity (gender, ethnicity, sexuality) tends to be overshadowed by a productionist or statist framework which centralises questions of redistribution and structural change, while in the postmodernist option structural factors as well as the commonalities among movements are swept aside in the flux and fray of contingency and difference. But can a comprehensive theory of politics be built exclusively on one of these two approaches?

Feminists such as Fraser have responded in the negative. Rejecting a monological view of politics, feminists tell us that to define the political in terms of production relations or state sovereignty or contingent identity relations is in itself a political manouevre which privileges *a priori* some relations over others. A more illuminating way of exploring politics, feminists argue, is to begin by identifying sites of oppression and resistance wherever they may manifest themselves.[46] Once this is done, Fraser reminds us that we need to take a closer look at the relationship among different types of political claims (deriving from gender, class, race, sexuality, religion and nationality). If we do, we will find a complex web of overlapping power relations that cut across class, ethnic, national and gender lines. The challenge for the post-positivist international relations theorist, then, is to identify in what ways these various power relations intersect and to examine how questions of political economy, state sovereignty, cultural marginalisation and cultural identity are interrelated. In short, the multiple and overlapping nature of power relations militates against one type of social movement being elevated above others; social movements, like other forms of politics, are multivalent and complex.

A further task for the post-positivists in international relations is to study the role of 'discourses' in the generation of social movements and the construction and politicisation of identities. For example, both the international indigenous peoples movement and the women's global network for reproductive rights movement became mobilised and gained momentum, in part, as a response to changes in international political discourse. In the former case, the publication of a 1972 United Nations (UN) report on 'indigenous peoples' played an important role in the development of the movement as it both identified indigenous peoples as sharing a collective experience and provided a vocabulary (accepted by the UN establishment) around which they could mobilise. In the latter case, the adoption of 'population discourses' on the part of UN agencies has given the women's reproductive rights movement considerable momentum and a theoretical and political focus.

Moreover, a sustained examination of the way in which solidarity and diversity can co-exist within movements could also make a contribution to advancing the Fourth Debate. If we take the work of Sylvester seriously and if we do our own historical and empirical research into the political practices of movements, we realise that there need not be an either/or choice between solidarity and diversity. Take for example the women's health movement. Feminist activists sought to create a united 'political front' which, on the one hand, allowed room for the diversity of the women's organisations involved in the movement, and on the other, enabled these women to act in unison when it came to intervening effectively in the UN Cairo Conference proceedings. This goal was achieved in 1994 with the creation of the *Women's Voices '94 Alliance*. While each member was permitted to undertake its own independent work on reproductive health matters and to lobby accordingly, in certain strategic moments the 'political front' took over as the dominant negotiating force. This strategy proved very effective in maintaining both diversity within the movement (since each organisation was granted a large degree of autonomy) and a focus on women's health issues during the preparatory meetings for the Cairo conference.[47] The politics of solidarity and that of difference, therefore, as some feminists have suggested, should be explored together rather than as mutually exclusive phenomena.

One last concern that feminist theorists have encouraged us to address is the question of political commitment in post-positivist international relations research. As we have seen, feminists argue that political commitment does not have to mean blind partisanship. On the contrary, precisely because the development and success of a movement depends to a considerable extent on a realistic and sober evaluation of its strengths and weaknesses, the committed scholar has the incentive to adopt a critical stance towards her subject matter. The result can be a more self-reflexive, less dogmatic and, in this sense, more objective scholarship. For those of us engaged in the Fourth Debate this may be a vitally important lesson. After all, one of the main aspirations of post-positivist international relations theorists is to construct a theory of politics that is more sensitive to the various forms of oppression and resistance in contemporary international life. In approaching this task, we would do well to remember Fraser's words:

> Think of them (intellectuals)... as occupying specifiable locations in social space rather than free-floating individuals who are beyond ideology... Think of them... as participants on several fronts in the struggles for cultural hegemony. Think of them also, alas, as mightily subject to delusions of grandeur and as needing to remain in close contact with their political comrades who are not intellectuals by profession in order to remain sane, level-headed and honest.[48]

CONCLUSION

The primary purpose of this chapter has been to bring to light the contributions that feminist theory can make both to an increasingly sterile debate among critical theorists and postmodernists as well as to the theorisation of politics in international relations. Exploring the 'no man's land' which lies between these two perspectives, feminists such as Fraser, Harding and Sylvester have pointed out the possibility of steering the discussion along a different path which affords fewer theoretical certainties but remains more open to the variable and complex task of theorising politics. The one guideline they do insist upon when exploring this relatively new theoretical terrain, however, is the necessity of grounding metatheory in social theory and social theory in the political practices of social movements.

NOTES

1. I would like to thank the editors for their helpful comments on this paper, as well as Mairi Johnson for introducing me to feminist theory in international relations.
2. See Michael Banks, 'The Inter-Paradigm Debate', in Margot Light and A.J.R. Groom (eds.), *International Relations: A Handbook of Current Theory* (London: Frances Pinter, 1985), pp. 7–26.
3. See Mark Hoffman, 'Critical theory and the inter-paradigm debate', in *Millennium: Journal of International Studies* (Vol. 16, No. 2, 1987), pp. 231–49; Hoffman, 'Conversations on critical theory in international relations theory', *Millennium: Journal of International Studies* (Vol. 17, No. 1, 1988), pp. 91–95. For his views on postmodernism and its role in this debate see N. J. Rengger and Mark Hoffman, 'Modernity, Postmodernism and International Relations', in Joe Doherty, Elspeth Graham and Mo Malek (eds.), *Postmodernism and the Social Sciences* (London: Macmillan Press, 1992), pp. 127–45.
4. The debate among these three groups has been called the 'post-positivist debate' by Steve Smith and Ken Booth. One scholar who has referred to it as the 'Fourth Debate' is Ole Waever, although his characterisation is quite different from my own. See Waever, 'The Rise and Fall of the Inter-paradigm Debate' in Steve Smith, Ken Booth and Marysia Zalewski (eds.), *International Theory: Positivism and Beyond* (Cambridge: Cambridge University Press, 1996), pp. 149–85. Although I do not claim that this Fourth Debate is disciplinary in scope – in fact many mainstream international relations scholars hardly know of its existence – it is considered very important for all those interested in defining alternative directions for the discipline. In fact Smith identifies this debate as the 'most important one for the future of international theory.' See Steve Smith, 'The Self-Images of a Discipline: A Genealogy of International Relations Theory', in Ken Booth and Steve Smith (eds.), *International Relations Today* (Philadelphia, PA: Pennsylvania State University Press, 1995), pp. 1–37.
5. By critical theory I am referring to a broad tradition of social theory which has as its central aims the revitalisation of Marx's thought along less economistic, scientific lines and the exploration of the role of consciousness in the activation of emancipatory political practice. Although this line of inquiry, on the whole, developed within the context of the Frankfurt School, I would argue that thinkers such as Gramsci can also be seen as contributing greatly to this tradition. In this way, I choose to neither reduce this tradition to the work of Habermas nor to generalise it to include all theories that propose emancipation as its central goal. For an overview of critical international relations theory see: Robert Cox, *Approaches to World Order* (Cambridge: Cambridge University Press, 1996); Hoffman, *op. cit.*; Andrew

Linklater, *Men and Citizens in the Theory of International Relations* (London: Macmillan Press, 1982); Linklater, *Beyond Realism and Marxism* (London: Macmillan Press, 1990); and Mark Neufeld, *The Restructuring of International Relations Theory* (Cambridge: Cambridge University Press, 1995).

6. In international relations the main proponent of this perspective is Andrew Linklater.

7. A sample of postmodern international relations works includes : David Campbell, *Writing Security: United States' Foreign Policy and the Politics of Identity* (Manchester: Manchester University Press, 1992); James Der Derian, *On Diplomacy: A Genealogy of Western Estrangement* (Oxford: Basil Blackwell, 1987); Michael Shapiro, *The Politics of Representation: Writing Practices in Biography, Photography and Policy Analysis* (Madison, WI: University of Wisconsin Press, 1987); R.B.J. Walker, 'Social Movements/World Politics', *Millennium: Journal of International Studies* (Vol. 23, No. 3, 1994), pp. 669–700; and Walker, *One World, Many Worlds: Struggles for a Just World Peace* (Boulder, CO: Lynne Rienner, 1988).

8. Walker, *ibid.*, p. 77.

9. One of the most interesting examinations of the debate between feminist critical theorists and postmodern feminists can be found in Seyla Benhabib, Judith Butler, Drucilla Cornell and Nancy Fraser, *Feminist Contentions: A Philosophical Exchange* (London: Routledge, 1995). See also Nancy Hirschmann and Christine Di Stefano (eds.), *Revisioning the Political: Feminist Reconstructions of Traditional Western Concepts in Western Political Theory* (Boulder, CO: Westview Press, 1996); Linda Nicholson (ed.), *Feminism/Postmodernism* (London: Routledge, 1990); and Stacey Young, *Changing the World: Discourse, Politics, and the Feminist Movement* (London: Routledge, 1997).

10. For an in-depth exploration of feminist ontology from both the postmodern and critical theory perspective see Kathy Ferguson, *The Man Question: Visions of Subjectivity in Feminist Theory* (Oxford: University of California Press, 1993).

11. Nancy Hartsock, 'The Feminist Standpoint: Developing the Ground for a Specifically Feminist Historical Materialism', in Linda Nicholson (ed.), *The Second Wave: A Reader in Feminist Theory* (London: Routledge, 1997), p. 229.

12. Seyla Benhabib, *The Situated Self: Gender, Community and Postmodernism in Contemporary Ethics* (New York, NY: Routledge, 1992) p. 5.

13. *Ibid.*, p. 339.

14. Benhabib, *op. cit.*, p. 12.

15. Jane Flax, 'Postmodernism and Gender Relations in Feminist Theory', in Nicholson, *op. cit.*, p. 56.

16. Seyla Benhabib, 'Feminism and Postmodernism: An Uneasy Alliance', in Benhabib, et. al., *op. cit.*, p. 23.
17. Judith Butler, 'For a Careful Reading', in Benhabib, et. al., *op. cit.*, p. 128.
18. *Ibid.*, p. 133.
19. Kathleen Lennon and Margaret Whitford (eds.), *Knowing the Difference: Feminist Perspectives in Epistemology* (London: Routledge, 1994), p. 7.
20. Sandra Harding, 'Feminism, Science and the Anti-Enlightenment Critiques' in Nicholson, *op. cit.*, p. 86.
21. *Ibid.*, p. 101.
22. This is not to say that the divisions between feminist critical theorists and postmodern feminists in international relations are not visible. They are visible, with the large majority of international relations feminist scholars belonging to the camp of foundationalist theorising and a much smaller minority aligning themselves with the postmodern camp.
23. See Christine Sylvester, *Feminist Theory and International Relations in a Postmodern Era* (Cambridge: Cambridge University Press, 1994); Sylvester, 'Empathetic Cooperation: A Feminist Method for IR', *Millennium: Journal of International Studies* (Vol. 23, No. 2, 1994), pp. 315–34; and Marysia Zalewski 'The Women/"Women" Question in International Relations' *Millennium: Journal of International Studies* (Vol. 23, No. 2, 1994), pp. 407–24.
24. Nancy Fraser, 'Pragmatism, Feminism and the Linguistic Turn', in Benhabib, et. al., *op. cit.*, pp. 157–72; and Fraser, 'False Antithesis', in Benhabib, et. al., *ibid.*, pp. 59–74.
25. Nancy Fraser, *Justice Interruptus: Critical Reflections on the 'Postsocialist' Condition* (London: Routledge, 1997), p. 208.
26. Fraser, *ibid.*, p. 166.
27. Nancy Fraser, *Unruly Practices: Power, Discourse and Gender in Contemporary Social Theory* (Minneapolis, MN: University of Minnesota Press, 1989), pp. 106–07.
28. Fraser, *Justice Interruptus, op. cit.*, p. 6.
29. *Ibid.*, p. 113.
30. *Ibid.*, p. 2.
31. Fraser makes it clear that she understands the terms 'political', 'economic' and 'domestic' as cultural classifications and ideological labels rather than structures, spheres, or things. See Fraser, *Unruly Practices, op. cit.*, p. 166.
32. *Ibid.*
33. Despite the strengths and insights of this conceptualisation of 'the political', there are inconsistencies. It is not entirely clear what Fraser means by 'the discursive-political'. Is a matter political simply because it becomes the subject of contestation among a range of discursive publics, as she suggests, or is it the fact that the matter being contested is *also* being

presented as the proper subject of public policy and public responsibility? If it is just the former, then any topic that is discussed and debated widely should be considered political no matter how trivial or unconnected it is to power relations or oppression. By this standard the life and times of former Princess Diana would be the most important political issue of the decade. If it is the latter, however, the question of the limits or boundaries of the political, in capitalist societies, would make more sense since it would refer to the way in which some issues are considered to fall within the jurisdiction of public policy and public responsibility, while others are relegated to the private sphere. This way of conceptualising the discursive-political would make Fraser's two forms of 'the political' much more interrelated than she suggests, since it is the desire for or fear of state intervention and public legislation (the official-political) that makes an issue discursive-political in the first place. In other words, although many of the new social movements today are uninterested in taking over the state *per se* or even participating in governmental politics, they are all concerned with obtaining recognition from the state and an acknowledgement that the issue around which they are mobilised is a matter of public concern and responsibility. Furthermore, at least in democratic societies, it is often only when an issue has gained the attention of the state, or agencies of the state, that it is brought into the public sphere via the media and gets serious public attention.

34. Fraser, *Unruly Practices, op. cit.*, p. 170.
35. One example of such a movement may be the recently formed Promise Keepers, a fervently Christian movement of men seeking to reestablish their dominant role within the family and as leaders of the faith. For a report on this movement see Ron Stodghill, 'God of our Fathers' in *TIME* Canadian Edition (Oct. 6, 1997), pp. 52–58.
36. *Ibid.*, p. 164.
37. Expert discourses emanate from institutions of knowledge production and utilisation such as universities and think tanks and from legal and other professional journals and associations. See Fraser, *Unruly Practices, op. cit.*, p. 173.
38. Fraser, *Justice Interruptus, op. cit.*, p. 12.
39. It may be argued at this point that Fraser too is working with a false dichotomy, that is, the contrast between the material world and the discursive world. Indeed, many postmodernists deny this distinction. I would argue however that her effort to maintain a separation between these two spheres is important precisely because the attempts of critical theorists and postmodernists to theorise both together has led to the prioritisation of one side over the other. Furthermore, I would suggest that it is important to remember that discourses and institutions have a different life span and that the former may change more rapidly than the latter. See Robert Cox, 'Social Forces, States and World Orders: Beyond International Relations Theory', in

Cox, *op. cit.*, pp. 85–123.

40. Linklater, 'Dialogue, Dialectic and Emancipation in International Relations at the End of the Post-War Age', *Millennium: Journal of International Studies* (Vol. 24, No. 1, 1994), p. 126.

41. Walker, *One World, Many Worlds*, *op. cit.*, p. 151.

42. See Cox, *Production, Power and World Order: Social Forces in the Making of History* (New York, NY: Columbia University Press, 1987) and Cox, *Approaches to World Order*, *op. cit.*

43. One interesting postmodernist effort to examine this question has been made by Warren Magnusson in his book *The Search for Political Space: Globalization, Social Movements and the Urban Political Experience* (Toronto: University of Toronto Press, 1996).

44. One example of a universalist application of communicative rationality can be found in Linklater's work. Not only does he take for granted states as the central unit of analysis of international relations, he also anthropomorphises them so that he can apply a Habermasian notion of moral learning to their foreign policies. In this way, one is left with the impression (although I do not think that this was his intention) that politics is limited to the realm of interstate interactions and that it is determined by a teleological dynamic which slowly but surely evacuates all conflicting power relations from the scene.

45. Discourses connected to the state and the economy would include, among other things, international law, international 'regimes', foreign policy, diplomacy and the political and economic ideologies of the dominant states. Discourses related to the cultural sphere include the secular and religious ideologies of nationalist and other social movements seeking to challenge these dominant discourses.

46. As Cynthia Enloe has pointed out, these sites may or may not be directly connected to the state or production and more often than not will appear at the margins of a social order rather than at its centre, that is, the state. See Cynthia Enloe, *The Morning After:Sexual Politics at the End of the Cold War* (Berkeley, CA: University of California Press, 1993) and Enloe, 'Margins, Silences and Bottom Rungs: How to Overcome the Underestimation of Power in the Study of International Relations' in Smith, Booth and Zalewski, *op. cit.*, pp. 186–202.

47. Loes Keysers, 'Is Gendering Population and Development Negotiations a Paradigm Shift? Alternative Body Building Exercises for Women's Reproductive Health Rights', p. 9. Paper presented at the 'World in Transition: Feminist Perspectives on International Relations Conference', June 14–16 1996, Lund University, Sweden.

48. Fraser, *Unruly Practices*, *op. cit.*, p. 108.

7. What is Left of the Domestic? A Reverse Angle View of Foreign Policy
Christopher Hill

In much writing on international relations over the last ten years, the concept of foreign policy has simply been by-passed. The trend of theoretical work has been determinedly away from the excitement over images, perceptions and cybernetics which brought foreign policy analysis to the fore in the 1960s and 1970s, and the metatheoretical interest in paradigms, emancipation and discourse has only occasionally collided with what has by now become the orthodoxy of a decision-making approach. On the empirical side, the impact of historical sociologists has both directed attention away from politics and insisted on a much longer perspective than that employed by those interested in what is done (and not done) by foreign policy-makers. Equally, the explosion of interest in international political economy, and its more recent offspring globalisation, have made the state itself, let alone its external arm, seem a puny thing, caught up in the seismic movements of the great forces of production, markets and technology.

This chapter seeks partially to redress the balance of the neglect of foreign policy by arguing that the concept in some form is indispensable if we are to grapple with the great imponderables of agency, responsibility and causation. In order to do this it begins with the basic distinction between the 'foreign' (abroad) and the 'domestic' (home) and examines how current writings on international relations tend to handle it. It goes on to outline six propositions constituting an argument for retaining the distinction as a central part of our understanding of the world, albeit in a constantly evolving form. A brief case-study is offered of one of the most illuminating collisions between domestic society and foreign policy of the post-war period, that of US involvement in Vietnam. The important challenges to the idea and practice of foreign policy represented by the three trends of globalisation, changing identities and regionalisation are then considered in turn, and the chapter concludes by focusing more explicitly on the issues at the empirical/normative border of foreign policy, namely democracy, legitimacy and responsibility.

THE FOREIGN-DOMESTIC DISTINCTION

The expanding universe of intellectual interest in international relations might have begun with a big bang of concerns about foreign policy, but it has moved relentlessly away from them ever since. Yet beneath the surface the picture is inevitably more complicated. It is difficult to avoid in practice the perennial human dilemma of how to assert our existential freedom in the face of what often seem overwhelming odds. This problem is often not addressed directly, but there are usually implicit positions taken on it. Where it is discussed, the old Great Man (sic) versus Great Impersonal Forces contest, is now reformulated in terms of 'Structure versus Agency'. This latter debate is important and vigorous, but most attention has been focused on disagreements about the nature of Structure, with Agency and where it might be sited revealingly neglected, in part through genuine uncertainty as to where agency does (and should) reside in the changing international system.[1]

Nonetheless, the lack of recent theoretical interest in foreign policy is complicated by the fact that in some respects the state has made a comeback. In approaches to the international system, neo-realists are only slowly retreating before the criticisms of liberals, the English School and various other kinds of reflectivists (most of whom are themselves hardly dismissive of the state). Indeed in the United States the continued dominance of the 'scientific' paradigm, now grafted powerfully onto the economists' tool of rational choice theory, has meant that the state has survived as the fundamental unit of analysis. After all, if you believe in measuring and calculating it is much easier to do so on the assumption of a known, tangible actor than on the basis of a kaleidoscope of signifiers and signifieds. At the other end of the ideological spectrum, by coming to terms with the state and international relations, Marxism has been able to survive – as an intellectual tradition at least – the devastating blows dealt by the events of 1989-91.[2] Even in an area where the state has been seen as outmoded from the beginning, that of European integration studies, there is no doubt that the enthusiasms of early federalists and functionalists have been tempered, first by neo-functionalists, and then more sharply by invading realists, neo-realists and liberal intergovernmentalists.

None of these groups, however, has had much to say about foreign policy as an activity, despite interesting work by individuals. The neo-realists are mostly interested in systemic change, while the neo-institutionalists tend to be preoccupied with games and bargains played out in international organisations. The English School is preoccupied with the notion of international society, and the ubiquitous decomposition of 'public policy' by pluralists into 'issue-areas' tends to marginalise the concept of a community thinking collectively about its relations with foreigners. Marxists tend to be interested, like other political economists, in relations between the state and the transnational processes of capitalism. Even the direct proponents of foreign policy analysis have either been driven back into contact with international historians, or have preferred to keep their heads down to avoid the frontal assault of those for whom foreign policy, and the domestic/foreign dichotomy out of which it arises, are simply anachronistic notions.[3]

This chapter resists the marginalisation of foreign policy by default, the facile (but ever more common) assumption that foreign policy analysis is a branch of realism, and that because the distinction between foreign and domestic is undoubtedly blurred, what states do to and with each other does not matter.[4] It confronts this tendency to overstatement by reversing the direction of analysis, by looking at the video from the opposite side, as it were. It accepts that the notion of the 'domestic' is inextricably bound up with that of foreign policy, and asks whether, if foreign policy is now so lacking in content, where the logic then takes us in regard to the idea of domestic society. Conversely, if 'domestic society' is not a self-deluding piece of conventionalism but a category with analytical and political meaning, what does that in turn imply for the meaning of foreign policy?

Some contemporary theorists, from various starting-points, are honest enough to agree that the domestic and the foreign are siamese twins, and to be willing to take the logic on to the point of dispensing with both. Thus Cynthia Weber argues that the 'domestic community' is simply not identifiable in practice: 'it is fruitless to attempt to locate a domestic community that can be said to be the sovereign authority in a state.' She cites Raymond Vernon, John Herz and Keohane and Nye as making it difficult 'to imagine not only boundaries but also "domestic" as opposed to "international" spaces – not to mention "communities".' Their evidence apparently shifts 'the focus of analysis away from questions such as "who is and is not a member of a particular domestic community?" to questions such as "how is it possible to speak of 'communities' at all?"'[5]

Weber, like R.B.J. Walker, whose book *Inside/Outside* she cites, is interested in language, and in the political 'fictions' and 'simulations' associated with key notions in international relations which they both hold to be more constructions than anything anchored in external referents. Walker says that his concern is

> to destabilise seemingly opposed categories by showing how they are at once mutually constitutive and yet always in the process of dissolving into each other. The nice straight – spatial – lines of demarcation between inside and outside or realism and idealism turn out to be shifting and treacherous.[6]

Later on, he points out that there are serious implications for 'the established routines of democratic theory and nationalist aspiration' once the principle of state sovereignty becomes accepted as contested and problematic: '[r]ather a lot hangs on whether the assumptions and silences of received wisdom and legal convention can remain relatively undisturbed, and on whether we can remain convinced that there is a here here as well as a there there.'[7] There is an important point being made here: if we cannot accept as any more than particular historical fabrications concepts like 'state', 'sovereignty', 'community' and 'foreign', then other central parts of our modernist world must also be shaken to their very foundations, namely our preoccupation with democracy and accountability on the one hand, and that with national identity and proximity on the other. The idealists must engage with the meanings and uses of sovereignty, just as the realists must accept that 'international relations already is an expression of ethical possibility.'[8]

It is important to note that it is not just the postmodern turn which has called into question the distinction between the foreign and the domestic. I have already cited those who have simply lost interest in serious questions of agency, but there remain those analysts who, while impatient with both traditional emphases on the state and the introspection of poststructuralism, are still committed to understanding action and responsibility. They insist that the human dimension of international politics must command our attention, and not be screened out by arcane academic concerns. There are two types of thinkers in this group, those normatively committed to an ethics of humanity rather than that of the ideal state, and those who are convinced that whatever we might prefer, the action of politics and economics has now broken free of the state onto the global stage, where it must be understood transnationally and holistically.

Ken Booth is a good representative of the first group, the ethical globalists. His powerful indictment of 'human wrongs' leads him to pitch the call for action at the level of the citizen of the world rather than at that of state citizens with their prioritised responsibilities to fellow-nationals, neighbours, allies or those to whom an historical debt is owed.[9] He wants a 'global moral science for the twenty-first century... more sociology than strategy, more macro-history than micro-diplomatic history and more paradigms than power politics.'[10] And there are indeed strong arguments for the position that suffering makes the same claims whether it takes place next door or in another hemisphere, and for the view that the existing order of things, the myopia of nations and states, produces much appalling suffering. Booth does not flag up the problems of choice, limited resources and priorities, the unromantic business of foreign policy, indeed of politics as a whole – except, interestingly, to subscribe to the orthodox position on the dangers of military interventions.[11] In his approach, actions must be primarily taken by concerned citizens if we are to move into a new era by transcending the institutionalised imperialism of the state system.[12]

If Booth is primarily an ethical globalist, he and his fellows are outnumbered by the empirical globalists, those who believe that the state's actual powers, like it or not, are shrinking relentlessly, and that the debate about policy and action has to take place at another level (which, however, is rarely clear). Jan Aart Scholte is a half-way house member of this second group, in that his values convince him of the value of a globalist approach just as much as his observations do.[13] But most members of the school are hard-headed people much more concerned with money and business than with human rights. It is now such a commonplace to argue that governments cannot manage their own economies, that the global eco-system is out of control, and that consumerism is homogenising us all, that even academics can talk without embarrassment about something called 'global governance, while not managing to specify who might be governing what.'[14] Politicians have made it the new realism; when Tony Blair swept to power in Britain in 1997 his programme eschewed serious economic change on the grounds that the 'global economy' was too big for a mere statelet of 56 million people to buck.[15]

It is, therefore, from a number of diverse points of view and with a variety of powerful arguments that the associated notions of foreign policy, domestic community, statehood, and sovereignty are becoming marginalised in contemporary political discourse. Even the preferred concept of agency, so much more neutral than that of 'policy', has been honoured more in the breach than the observance, leaving the agency-structure problem in a seriously asymmetrical condition.

And yet what has been put in the place of the foreign/domestic divide? Robert Keohane and Joseph Nye seemed on the road towards supplanting it with their exegesis of transnationalism in the 1970s, but this turned out to be more a refinement of the basic model than a challenge to it, in that their next and best-remembered book concerned itself with how states manage interdependence, not how they were being by-passed.[16] Susan Strange has made the most systematic attempt to construct an alternative general model for world politics with her four authority structures of money, labour, production and knowledge, where 'states and markets' compete in varying ways according to their sources of power.[17] Much of Strange's analysis is persuasive, but even on her own logic, there remains the problem of how big is the residual category into which fall what she calls 'security' questions – that is, conflicts like those between Iran and Iraq and in Bosnia, or rivalries such as those of Greece and Turkey over Cyprus. Some category, some residuum....

The ineluctable tendency of those who wish to relegate the state to a position of lesser importance in international relations is that of downgrading politics. This is often not intentional; those falling into Martin Wight's 'revolutionist' category, after all, might see the politics of the international class struggle, of centre-periphery relations, or between clashing civilisations as rather more substantial than the politics of conventional inter-state relations. But generalisations of such a high order of magnitude are difficult to cash in, in terms of cases, choices and mobilisable political forces. They tend to remain at the systemic level, with a strong element of determinism and the identity of the central actors never being quite clear. The idealists, who make up one part of this group, are by definition calling for a new kind of politics and it is hardly fair to expect them to describe in detail what it might look like. Still, it is clear that just as the international law and institutions school of the 1920s became increasingly remote from the compelling political action arising from the rise of Nazism, so the human rights-plus-environment agenda of the ethical globalists can find it difficult to go beyond hand-wringing from a distance. Although it is true, as Steve Smith has argued, that their central point is to challenge the conventional view of politics (i.e. as what policy-makers do), it remains true that if you inherently distrust, or demote the importance of states and governments, it is then difficult to give convincing guidance on how matters are to be carried forward in the face of such dangerous problems of foreign policy as China's relations with Taiwan or the Arab-Israel dispute.[18] Problems of this kind – and no-one could dispute their significance for the lives of millions – in practice require a lead from states, combining amongst themselves and with other actors, if any progress is to be made.

Even on the agenda of empirical globalism, whose central point is that realism now requires a focus on markets not states, it is often difficult to see where the politics – in the sense of conflict over the authoritative allocation of values – has shifted to. The fate of planet earth, for example, is often presented either as a matter of individuals changing their habits of consumption or as an apocalyptical revelation which even spiritual renewal can hardly prevent.[19] Old-fashioned political quarrels, like that between Britain and Norway over acid rain, are mere footnotes to the main text. Similarly, in relation to the power of international financial markets and the global structure of production, the common implication is that neither states nor any other political entity have any real business in intervening – if they do, the consequences will be damaging to efficiency and quite probably counter-productive. This is indeed the trend of events in western Europe, with a deliberate attempt being made to ring-fence the single market and Central Bank from political interference.

Overlooking politics in relation to the environment or international economics will not, however, take understanding very far. The only way to address pressing ecological issues is to convene summits such as those at Rio and Kyoto, with the hope of agreeing conventions such as that of Montreal in 1987 on ozone-depleting chemical substances.

And these meetings are hot with politics, as arguments rage over the definition of the problem, responsibility for change and the distribution of costs. Moreover states are still the main actors at these meetings, which are intergovernmental affairs where the extensive, crucial, NGO activity necessarily focuses on obtaining changes in official policy. Unless major states like Brazil, China and the United States can be persuaded to act, the whole environmental enterprise will founder. Insofar as 'solutions' are possible to the ecological problems the world faces, the answer lies in the hands of us all as inhabitants of the globe; but only the most optimistic believe that effective actions can be mobilised by 'thinking globally and acting locally' – that is, by doing without governments and the politics of influencing governments.[20]

The same is true on the general economic front. It would be a bold observer, for example, who would argue that the EU will settle down into a smooth, self-administering economic system, with only minor political disagreements. *A fortiori*, the *politics* of globalisation will continue vigorously, in a multi-level competition involving states, blocs, regions, companies, trades unions and pressure-groups. There are indeed complex processes at work here which are difficult to conceptualise, but whether we do it by issue-area (trade, money, business,) or by region (Asia-Pacific, NAFTA, Europe) the further the analysis is pushed the more evident it becomes both that politics is central to outcomes and that government policies are a major source of the politics in question. The history of oil production in the Gulf and the current competition over oil and gas exploration in the Caspian Sea are cases in point.

Thus although it is common both to assume that states are of decreasing importance and to relegate politics to the margins of the analysis, neither tendency is justifiable. Indeed, the two should be delinked, for international politics will continue to be important whatever the fate of the nation-state, at every point along the way short of world government. Conversely the state is about more – and less – than politics, since it performs symbolic, constitutional, legal and administrative functions which are affected by politics but far from being co-terminous with it.

The ideas of politics and agency, however, should be kept in harness. To posit agency in international relations is to imply politics, and to posit politics is to imply agency. Unless one adopts the eccentric position that agency is of little consequence, space has to be found for politics. Because the narrow perspectives of realism and power politics may have held too much sway in the past, this is no excuse for its current use as a straw man while no serious attempt is made to reconceptualise agency. War is changing, but it is not over. The security agenda has widened, but this means only that new issues, like crime and migration, have become politicised internationally, not that functionalism is in the ascendant. Theories of world politics always have to take into account great systemic changes like the 'big bang' of financial services, or the collapse of the Warsaw Pact, but that does not mean that they may neglect to explain where agency might be located or the implications for the nature of political argument.

This is particularly true given that few members of any school in international relations find themselves able in practice to do without some kind of concept of domestic society, and therefore some form of separation of humanity into distinctive units. Scholte wishes eventually 'to dispense with even the analytical separation of the domestic and the international'[21], but recognises the difficulties. He has already conceded that 'although the process of transformation unfolds in a world-societal context... social change does not as a result take identical forms in each region, country and locality.'[22] Equally, those who argue that inter-state violence is now anachronistic (how long have liberals been promulgating this view?) still need the state as the base-line category which enables them – as with the democratic peace debate – to compare degrees of anarchical recidivism.[23]

SIX PROPOSITIONS AND AN ARGUMENT

It is now time to set out more explicitly the argument for a continued usefulness of the domestic/foreign division, starting with the notion of domestic society.

This chapter is not intended as a simple defence of the idea of domestic society and therefore of that of foreign policy. That would be to fall back on the polarities and pendulum effects that characterise too much of what passes for debate in the social sciences. Rather, it starts from the assumption that there are indeed important criticisms to be made of the more tyrannical definitions of foreign policy, and that the case for shifting our subject matter from inter-state relations to 'world politics' is a substantial one. There is little point in trying to resuscitate realism or the comatose notion of 'national interest'. These ideas are largely played out. On the other hand, it is also insufficient to create a new antinomy between world politics and domestic policy, as many contemporary writers do by default, unwilling as they are to extend the transnational logic so far as abandoning the notion of the domestic altogether. Most observers instinctively feel, even where they are critical of the ideas of state sovereignty and national security, that palpable domestic communities still exist. If this is the case, however, then we need conceptual tools to analyse how relations between these communities are, and should be mediated. The space between them varies over time and space, but it is usually evident, and foreign policy still does a good deal of work within it.

The justification for this view is based on six inter-linked propositions, which are set out briefly below. The propositions are then elaborated more obliquely through a case-study, that of the United States in Vietnam, which illustrates the complex but substantive character of the interaction between domestic and foreign variables.

1. *The domestic realm has a real meaning* – always constitutional and political (unless the state has completely broken down), and usually also economic, cultural and symbolic. People still experience directly both the constraints and the advantages of their own society, despite the evident tendencies in advanced societies towards a virtual and ultimately sentimental sense of community created by television and the heritage industry. Taxes, the provision of services (or lack of them) and the nature of policing all make the state of immediate importance to its citizens, who look instinctively for its protection when trouble hits them while abroad. Likewise, their sense of living in a distinctive environment is likely to be strong even where nationalism is not rampant. Historical experience, passed on by families, churches and literature, and still present in language, food, music and architecture, all work to create a sense of belonging which by definition also involves a sense of 'not belonging' elsewhere. It is true that some people feel that they do not belong in their own society, but this has always been true historically, and territorial identifications are far from extinct.

The United States, for example, may be the source of global capitalism, but Americans bow to no-one in their ability to distinguish between home and abroad. Moreover the translation of the sense of the domestic outwards from family and neighbourhood to society and polity has probably gone further in our own time than ever before, now social mobility has undermined traditional attachments to 'pays', 'paese' and 'heimat'. Even in new and/or badly functioning states this division is apprehended, paradoxically by understanding how much better life might be outside the boundaries within which citizens are confined, be they Burmese or Kurdish, Albanian or (in the 1980s) Vietnamese.

2. *The idea of society is interdependent with that of the domestic* – in the sense that once we assume that people live in distinguishable societies we also posit the internal relations of that society and the exclusion of others who live outside it (all societies have immigration policies, however partial and relaxed). This inclusion/exclusion may in some cases be more contractual than affective, but this does not mean that we need allow the relations among the contracted no distinctive quality.[24] In other words, any group of human beings which conceives of itself as living in the same society must also operate on the basis of a distinction between the domestic and the foreign.

Conversely, if a group has a strong sense of what is foreign (for example, Catalan attitudes towards the Spanish) then it may be halfway towards constituting a separate society. That a perception of foreignness is not sufficient in itself to justify the attribution of the term 'society' is because of the role of the state. The Spanish state has been long enough in existence to have created an integrated Spanish society over and above those which might exist in the regions. The relationship between state and society is a difficult problem, which also exists with regard to states and nations. The three phenomena are deeply interpenetrated but still distinct from each other. All that can be said in this context is that while the existence of a state does not guarantee the existence of a coterminous society, it is very difficult for a sub-state society to be sustained if the state structure does not encourage it, and virtually impossible for a transnational society to exist across state borders without provoking the tensions which destroyed East Pakistan in 1971 and have wreaked havoc in Kurdistan, Northen Ireland and Yugoslavia. In all these examples, the relationship between the state and what had come to be considered the domestic realm was disastrously out of kilter.

3. *The domestic realm nonetheless cannot be sharply demarcated*, any more than individuals can wholly separate out their private (domestic) and public (social) lives. The domestic environment will never be impermeable, however hard the isolationists or autarkists try. This point helps to clarify the relationship between the state and the notion of the domestic. The state has a clear frontier, both literally and constitutionally. Its power, even its capacity to exercise sovereignty may often be restricted, but where its legal authority starts and finishes should always be clear. If not, then its very statehood will be in question. The domestic realm, by contrast, has fuzzy edges. It may extend to include diasporas, in the way that Jews world-wide may feel they have common cause with Israel; it will probably change over time, in the way that white and black South Africans are coming to see each other as part of each other's *home*land, after decades of 'othering'.

Thus 'domestic' and 'foreign' are two ends of a continuum rather than polar opposites. Each end point represents a distinctive set of relationships – just as some items on a nightly news broadcast are of interest only to the country concerned, and others are shared across the globe – but towards the mid-point it will be difficult to distinguish the two, in terms of scope, responsibility and action. To any society some events on other parts of the planet are bound to feel remote, even alien. Only the greatest dramas or tragedies will touch the nerve of human empathy we all contain within us for the plight of any other fellow-being. Conversely, only the most puritanical show no interest in their neighbours' affairs, and this principle of vicinity is extended through common language, mass media and sport to the millions of others who share the same sense of the domestic, usually in the form of a state.[25]

4. *Society and state*, the bedrocks of the domestic, *are the primary but not exclusive source of **community**.*[26] They must not be confused with the notion of community any more than with each other. A successful state or society needs to be built on working communities, but it is difficult for the closeness inherent in a community (usually defined as the people living in one locality, or as having cultural or religious characteristics in common) to extend across a whole state.[27] In practice we tend to live simultaneously in various different communities: the family, a church, perhaps an ethnic minority, a profession, a consciousness (of class, shared experiences, moral priorities or whatever), even a generation. Each of these (depending on their strength) may have their own internal/external (domestic/foreign) dimensions. Such multiple communities complicate the operations of the state and mean that to some degree international politics has a 'mixed actor' quality.[28] If foreign policy does not adapt accordingly, it risks becoming irrelevant. Usually, if belatedly, it does adapt, as we can see with the involvement of many more government departments than the foreign ministry, and the growing recognition of embassies that their work now requires them to go out beyond government circles into the wider society of their host country. In the final analysis, however, communities do not have the same capacity as a state or an overarching society to generate the domestic/foreign divide, held back as they are by lack of either size or clear territorial demarcation. When loyalties clash it is a serious disadvantage to be caught in the no man's land between integration in domestic society and foreign policy actorness. The Palestinians expelled from Jordan and Lebanon, or the modernising elites in Turkey, excluded from the EU they yearn for, have discovered this lesson by painful experience.

5. Action and agency may be possible beyond the confines of the state, but there is as yet *no clear way to organise democratic accountability without falling back on the formations of an organised polity and of society*, and on the *right* of a people to make its own decisions on its own fate. This is one of the

main dilemmas of modern international politics, and indeed of international theory: how to reconcile the evident facts of interdependence, multilateralism, collective approaches to security and the rise of democracy, with the absence of any legitimate means of allowing the people to hold international policy-makers to account. Within the European Union, the only serious supranational entity, the inadequacy of both national parliaments and the European Parliament has produced concern (but little action) about the 'democratic deficit', and there has been some talk about the need to 'democratise' international institutions. But it remains far from clear how the former could be dealt with short of federalism, and how the latter might be attempted in any way beyond weighting the votes available to governments (sic) on the basis of population. New escape-routes may well be found from this impasse, but for the foreseeable future citizens concerned about international affairs know that their main recourse is through their own government. Active international non-governmental organisations (INGOs), such as Greenpeace, to some extent transcend this limitation by stirring up domestic opposition on a transnational basis. But their performance is patchy and is hardly a substitute for legitimate democratic control.

6. *Foreign policy, ergo, remains a useful category* – to the extent that we believe a self-conscious society/polity exists, with its own decision-making system. Foreign policy represents the ways in which a domestic unit deals with other such units whose perceptions and interests differ, as they will even where underlying values are shared. The outside world will need coping with in terms of its intermittent acts of deliberate intrusiveness ('intervention' is the usual IR term) but there will also be a wealth of regulatory and distributional politics between states, as over the Non-Proliferation Treaty, where the French position differed from that of Sweden, and India's from that of Sri Lanka. Foreign policy is also about the various ways in which states (and other actors) try to shape the overall milieu in which all have to coexist.

The practicability of any given foreign policy is another matter, but we are doomed to have one even if we transcend the state regionally, or break it up from within – all that varies is the size of the unit with a foreign policy and what falls within its boundaries. If we have something recognisably domestic, that is, something which is primarily if not perfectly 'ours', then we must also have a foreign environment to cope with, however amenable it may be. The term 'foreign', like 'alien', has come to have threatening overtones, but it really refers only to difference and unfamiliarity.[29] In that respect it still has great utility in international politics, for even if the strangeness of strangers is becoming gradually attenuated, there are still many differences and senses of the 'other' built into our world. And only through the reverse angle of knowing our own domestic selves can we really understand what is 'foreign', and how to formulate our political choices.

It is really the word 'policy' which causes most of the problems in the term 'foreign policy'. It is almost as difficult to have an overall, coherent foreign policy as it would be to have a single 'domestic policy'.[30] On some occasions, usually after great set-backs or coming up to elections, governments make statements of some overarching 'foreign policy'. Even more rarely, they may announce new doctrines or objectives which amount to major breaks with the past. The Truman Doctrine was a classic example, and the current Labour Government in Britain says that its attitude towards the European Union is another. Most of the time governments run foreign policies in the plural, with only intermittent consistency between them. For most states there is still an ever-present flow of foreign policy *decisions* on a wide range of geographical and functional issues. These will not, in the late twentieth century, always be the province of the foreign ministry, and they will often fall in the grey area between the domestic and the international. But there will be enough problems with neighbours, expatriates, trade, defence, international institutions and so on to make foreign policy into a substantial concern for most states. This is a point easy to miss from the 'intermestic' vantage-point of EU politics.

THE UNITED STATES AND THE VIETNAM WAR

The relationship between the domestic and the foreign, interpenetrated, often confused but still meaningful, may be illustrated by a brief discussion of one of the most traumatic cases of post-World War Two international relations, the United States involvement in Vietnam. Although the Vietnam War supposedly occurred before the era of globalisation and postmodernity, the Gulf crisis of 1991 showed that the world's greatest power can still easily find itself fighting faraway wars against smaller adversaries of a very different character. Over Kuwait the outcome was swift, decisive and popular at home, but things could have gone badly wrong, and they might again in the future. Vietnam still has things to tell us, not least because sufficient time has now passed for a formidable body of scholarship on the subject to have built up.[31]

In this context what matters is the way in which the distinction between foreign policy and domestic society is fundamental to our understanding of US policy – although it was even more significant for the Vietnamese – and in two different ways. First is the familiar story of how military perceptions increasingly diverged from the way the war was seen at home. By 1968, this led to growing pressure on the U.S. Administration to back away from an open-ended commitment and eventually to accept that the war could not be won. Public opinion did not, as is popularly supposed, defeat the US in Vietnam by itself, but it did gradually begin to pull in a different direction from the logic of the war as seen by field commanders and indeed by the Oval Office.[32] That was, after all, one very good reason why the bombing of Cambodia from 1969 had to be kept secret. The ultimate success of the anti-war campaign led in time to a stab in the back myth not unlike that of the German military in the 1920s, even if a paradoxical twist was that the disabused ordinary soldiers, who knew the real horror of the conflict and soon became cynical about it, also felt that American society had not understood or appreciated their sacrifices overseas.[33]

The second aspect of the Vietnam War which illustrates the importance of the domestic/foreign distinction is the way in which President Lyndon Johnson tried to pursue two major policy goals simultaneously: one, the 'Great Society' programme at home, the other, victory in Vietnam. There was thus a seeming contradiction between the President's far-sighted and compassionate attempt to advance social justice and defeat racism in his own society, and his brutal prosecution of a technological war against Asian peasants whom he often abused as 'gooks'. At one level this simply seems to bear out the conventional realist picture of international relations requiring a very different approach from that of domestic politics. But that is not the argument being made here. There have been plenty of foreign policy-makers who have not behaved in the 'split' manner of a Johnson, a Nixon or an Andreotti (these last two being the reverse, having reputations for responsibility abroad and deviousness at home). The problem is rather that of the double gaps which exist between the priorities of domestic and foreign policies on the one hand, and the cultures of separate societies engaged in conflict on the other. In terms of priorities, Johnson saw his Great Society dream and the war as serving the same basic goal – the protection of American democracy. In practice his financial and political resources together produced a zero-sum situation, and the hawkish foreign policy, partly pursued to secure his flanks against conservative critics of Great Society, ended by undermining his whole Presidency. Because of the great but distinct demands of the two theatres, he ended by succeeding in neither. This was compounded by the chasm between the two societies, American and Vietnamese, with which he was struggling simultaneously. The US Administration's many miscalculations were often due to a simple failure to understand that American values were not naturally translatable abroad. As Johnson said plaintively, foreigners were just 'not like the folks you were reared with.'[34] Thirty years later, it is difficult to imagine that Bill Clinton, even with his greater personal sophistication, is much nearer to understanding the Iraqi or Iranian equivalents of Johnson's Vietcong. In a world where millions still think in terms of 'great Satans' or 'mad mullahs', it is premature to announce the death of domestic society.

CHALLENGES: GLOBALISATION, IDENTITY AND REGIONALISM

The six propositions and one case-study set out above have attempted to establish the argument for the continuing importance of the concept of foreign policy, while at the same time acknowledging the changes which have complicated world politics since 1945. Among these changes there are three major developments which now need more detailed consideration: globalisation, or the emergence of system-dominant global economic integration; changing identities and multiple loyalties; and regionalism.

Globalisation

Let us assume that globalisation represents a significant trend towards the creation of a world political economy. As a process it consists of a number of different elements, amongst which the most prominent are the liberalisation of trade and capital movements, the acceleration of technological change, particularly in the area of infomatics, the triumph of capitalism over communism and the increased mobility of the factors of production. What does this mean for individual societies and for the idea of the domestic?

Susan Strange has endorsed the view that we are living in a new mediaevalism, with authority sources dispersed and disconnected from the territory which has been the basis of the nation-state. States now have to share power with private business, professional associations and various kinds of international bureaucracy in a wholly inconsistent pattern.[35] Yet even here, where the states-system is characterised as barely relevant to anything, Strange makes a number of significant concessions, viz: that wealth requires the production of public goods; that it is difficult for transnational enterprises to provide many of these goods (despite the functionalist tinge of her argument); that it is a puzzle as to where authority might have drained away to, if it is indeed lost to the state (an echo here of the problems with John Burton's work[36]); that despite these new challenges some states are quite successful not only in providing their own public goods, but also in moulding the shape of the global political economy. In other words, there exists a *variety* of states, including the puzzling development of the recognition and survival of many micro-states. Some states choose a minimalist regulatory role, others a maximalist interventionism.[37]

So, where does this variety come from? No doubt it is partly dictated by the structuralist logic of the world political economy itself. But this leaves plenty of room for history, territoriality and acts of volition on the part of states. Here and there they derive from omnipotent dictators or juntas, but most of the time action is the product of political forces routinely interacting within the state. Furthermore, what if governments do rely on transnational elements to perform services they cannot supply, or do struggle to control their own economic fate? States have always used international service providers, from Florentine financiers to Swiss mercenaries, from Dutch missionaries to English railway engineers.[38] Moreover, the periods in history when governments have been able to deliver on their economic promises have been brief indeed, while if frustrations ever do mount to a politically intolerable level, reactions tend to set in, in the form either of inter-state cooperation or autarkic grab-backs of power. These have costs in terms of wealth, but may also produce gains in terms of self-confidence. As Geoff Mulgan has pointed out, it is important not to confuse 'the obsolescence of particular tools of government with the obsolescence of government itself.'[39]

In any case, we all know that firms and functional associations, whatever solace they provide for rootless cosmopolitans, cannot do more than scratch the surface of the things which most people really care about – jobs, security, education, welfare – because they do not exist to provide them. Politicians often fail on this count too, but at least their rationale is to try, and to ensure the conditions in which a community can generate their own products for the rest. Large companies can sponsor young artists' shows for the benefit of brand awareness. What they cannot do – any more than a factory robot can sing in the bath – is to create all the myriad trivialities which make up culture, and therefore our sense of who we are. These grow out of social interactions, and the communities embedded within them.

Too many people who write about international relations underestimate the continued role of communities. The 'global business civilisation' undoubtedly exists; it has replaced French country hotels with anonymous motel chains and it has made Anglo-Saxon rock music the basis of youth culture world-wide. But it is also an oxymoron. People are too resilient to allow their sense of community to atrophy completely, and the international system does not allow for more than a marginal mobility of labour, even if most people wanted to uproot themselves continually, which they do not.[40] We are therefore likely to continue for some time to compete – or not – in the world division of labour, on the basis of the same kind of messy, asymmetrical territorialism that has characterised most of recorded history.[41]

Changing identities

In discussing these constraints to mobility we are already moving into the area of the second challenge to the foreign/domestic divide, that of changing identities. It is now commonly argued, and not just by the postmodernists whose central point it can be said to be, that human experience at the end of the twentieth century is a matter of flexible identity and multiple loyalties, of resisting the political pressures to define the self in terms of nationality, so as to extend the sense of both sympathy and common interest well beyond state boundaries, while conversely denying a necessary commonality with those who live in the same territorial unit.

This perspective has its merits. Even if most people tend to stay put in one society and barely notice the changes of ownership of their multinational employers, there are good reasons why they may experience dissonance as a result of being pulled between different kinds of loyalty:

- New technology in transport and information technology has shifted the relationship between space and time to the point where millions of people can interact with former or potential enemies and acquire an interest in continuous relations, unbroken by war. What is more, national defence now seems less feasible than multilateral security systems.

- Citizens' normative concerns are taking on an increasingly global vision, with a scale, frequency and depth of penetration into society which is new. It can even be argued that on international issues a given society's domestic public opinion is now only likely to have much impact on its government's policy if it can elicit the support of what is often called 'world opinion'.[42]

- Many societies are multicultural, and therefore face a difficult balancing-act between inclusion and interference, between loyalty and breadth of human vision. In states under strain it is not uncommon to become dissatisfied with one's own state almost to the point of disloyalty. Internal 'others' can be more powerful than external bogies, as with Polish Communist Party in the 1980s or Yitzak Rabin in Israel. Often the only way forward seems through some kind of rainbow coalition which crosses the sacred frontier of national loyalty and may weaken its whole territorial basis.[43]

- The advance of the private sector at the expense of the *res publica* cuts across social cohesion. The recent victories of the free market principle have promoted the values of individualism, competitiveness and self-reliance, all of which tend to increase a sense of anomie.

- The experience of two World Wars has led many into an anti-nationalist reaction which has created inhibitions over a sense of self. This was particularly true of Germany and Japan, but in general nationalism is not an acceptable discourse in international institutions (which is not to say that it is absent). The language of diplomacy is therefore to a degree schizophrenic.

Yet this is where we reach the point of weakness of the case for identities becoming multiple, impermanent and non-territorial: it is an excessive generalisation. The empirical picture is highly varied, even contradictory. So far as nationalism is concerned, for example, even before 1989 and the overthrow of Soviet rule in eastern Europe, there were many places in the world where the desire for self-determination, or for nation-building, had produced a strong commitment to the nation as the superordinate political principle. Nicaragua, India and Malaysia are cases that come to mind, showing that the outlook of western Europe is hardly a weather-vane for the international community as a whole. After 1989, of course, this point was brought close to home for the Europeans: the anti-nationalism of Germany coincides with the constitutional nationalism of Scotland, the populist nationalism of Umberto Bossi's Northern League and the virulent state nationalism of Croatia and Serbia – to say nothing of the insurrections of ETA or the subtle forms of nationalism on display in Norway and Switzerland, both of which have opted out of the EU adventure.

This variety demonstrates in itself that the notion of a domestic community has not been rendered redundant by swirling currents of supra-societal, identity-dissolving postmodernity. Nationalism is resurgent and a strong sense of separate identity permeates political life in many states, whether deriving from top-down or bottom-up processes. Some states, like the Federal Republic of Germany and Brazil, have been gradually building the sense of unity and confidence which enables them to express distinctive interests inside alliances or international organisations, while some provinces of existing states wish to achieve statehood in their own right. Statehood and nationhood are hardly the only source of human identity, but for most people, except those living under conditions of oppression, the links between them are unproblematic. Social and national identity is neither trivial nor totalitarian, it is simply the orthodoxy.

In consequence, the tendency to find an 'other' more within the state than outside is still a minority tendency, while national establishments retain powerful instruments of manipulation over public attitudes to foreign 'threats' even in established liberal states. On the positive side of the coin, although many more citizens are concerned with human rights within other communities, and indeed give financial help on a regular basis to people they will never even see, foreign aid is also the most vulnerable part of a government's budget to domestic political pressures, and there are strict limits to what voters are willing to sacrifice in order to help developing economies (witness French opposition to Polish agricultural exports into the EU) or sustain sanctions against foreign tyrants. World opinion, moreover, remains an unpredictable, evanescent force in comparison to inter-state politics. The recent Taliban excesses against Afghan women were only highlighted so promptly because EU Commissioner Emma Bonino used her position to launch an appeal, while Italy and Russia tried to get the matter on the agenda of the UN Security Council.[44]

Finally, although the klondyke capitalism of the last 15 years has certainly damaged much of the social fabric in developed countries, and risks imposing intolerable strains on developing ones, it has not yet succeeded in rendering 'society' redundant in the way that Mrs. Thatcher famously observed. Indeed, in the US and now Britain it has produced the reaction in favour of Amitai Etzioni's 'communitarianism', and in the rest of the EU the Social Chapter to go alongside the Single Market. Furthermore in the more successful Newly Industrialising Countries of South-East Asia, such as Singapore, Malaysia and Indonesia, a strong social morality has been stimulated by government in a deliberate attempt to counter the more decadent effects of rapid growth.[45] This is to say, therefore, that the very real challenges posed by global capitalism to the inherent parochialisms of distinctive societies usually bring forth countervailing tendencies, as if the political need for some kind of community and agency were indestructible. As Friedrich Kratochwil points out in this book, identity and authenticity are the twin problems of modernity.[46] The popular hunger for them, and the resultant sense of the domestic, can hardly be regarded as atavistic.

Regionalism

The most overt threat to any particular domestic society and its associated foreign policy is that posed by regional state-building, which is why Eurosceptics are so sensitive to the prospect of federal advance in the EU. Federalism being a fundamental, constitutional change, unlike those socio-political trends discussed so far, it promises an irrevocable shift, an expunging of the past more radical than that represented by technology or free trade. The latter may actually represent a far more stealthy and final transformation, but at least it leaves the forms of political independence intact. Yet federalism of the kind immanent in the EU is uncommon in international relations; it is much more likely to develop as a solution to a particular state's internal difficulties, as could happen in Italy. In such a case the domestic would be accentuated – and multiplied – rather than diminished. In any event the EU is unlikely to make the step-change to a genuinely federal system in the near future; even if it did, its own subsidiarity principle is supposed to preserve the communities of the constituent states. It is true that these states would lose their legal personality in international relations, and that their national foreign policies and diplomatic systems would then disappear, but they would be replaced by one very large and potentially important foreign policy which no one could deny would be worth studying.

More interesting as a challenge to the domestic than federalism is something less structured, that is, the general tendency to aggregate and align states on a regional basis, which we may call 'regionalism'. Even if nationalism is in the ascendant, and areas like Padania, the Basque country or Scotland seek independence from their existing metropoles, they do not want full statehood. Rather, they want to renegotiate their external relations by becoming part of a larger 'Europe des régions', which need not be fully federal. They do not seek a wholly independent foreign policy, although they do want more freedom for their domestic community. Here we see that the argument cuts both ways. Although there is not an unbreakable link between the idea of domestic community and that of foreign policy, their delinking in an organisation like the EU destroys neither – it just shifts the level. The Northern League and the Scottish National Party envisage (although they may be somewhat naive in this) a more distinctive Padanian or Scottish identity in the world, sheltering under the wing of a pan-European foreign and defence policy. Their relations with their old metropoles inside the new Europe are to become looser, and more systematic with other similar regions. The 'domestic' then has two levels, local and intra-EU, while the 'foreign' is handled formally only by the federal entity, even if the regions may engage in extensive international relations of their own, on a 'low politics' basis.[47] This is all very well so long as the external environment remains conveniently unthreatening.

The EU is the major regionalist challenge to the foreign/domestic distinction precisely because it currently exists in a twilight zone between federalism and standard intergovernmentalism, thus blurring our conventional categories. By contrast, the other potential regional actors have hardly started down the same road. Organisations like ASEAN or the Contadora Group have seen the advantages of coordinating national foreign policies into what may be called 'diplomatic alliances', but they are strictly intergovernmental in form and sometimes only active on particular issues – as were the Front-Line States over South Africa. What is more, they rarely have much economic integration behind them in the way that the EU does, producing pressures for 'consistency' between a trade policy and traditional diplomacy. Conversely, where other regions have made progress with economic integration, they do not seem to have the wherewithal to produce a common foreign policy, witness Mercosur and NAFTA.[48]

Either way, the regional collaboration that is certainly on the increase in particular issue-areas, notably trade, is very far from the kind of integration that was envisaged back in the 1960s, when both neo-functionalism and regionalism were vibrant areas of international relations research. It barely scratches the surface of the various separate societies that engage in it. Even in Canada (to say nothing of Quebec), which has long been used to economic and defence interdependence with the United States, there exists a sharp sense of separate purpose which is precisely expressed through foreign policy. Ottawa places importance on the UN and the Commonwealth partly for reasons of identity arising from domestic pressures.

Thus inter-state cooperation does not necessarily undermine the sense of domestic difference, and even where foreign policies are coordinated, they are not *merged*. Indeed, the consequence of regionalism so far has been to make it more important, not less, to analyse international relations in terms of the interactions between national concerns and multilateral projects.

CONCLUSIONS

To recapitulate: the major challenges coming from globalisation, from the shifting foundations of identity and from federalist trends, have not rendered the idea and practice of a domestic society redundant. If it does not follow that a domestic society always means a distinctive foreign policy – think of Luxembourg – then the very existence of separate societies certainly promotes a realm of external relations, in which a different kind of politics may or may not occur, depending on conditions. The external consists in the desire of each society to have things which are its own, and which therefore necessarily excludes non-members. This includes the mechanisms of democratic accountability, which only work on a manageable scale and within a painfully established particular context. In this respect society is the more crucial unit than state, for a state without a functioning society (Lebanon in the 1980s) is not capable of projecting itself in international relations; there is simply no agreement on 'self' and no way of organising self-expression. Whereas an identifiable society can have a foreign policy of sorts even without the formidable advantages of full and recognised statehood (Taiwan; Turkish Northern Cyprus).

There is no disguising that there are problems at both operational and theoretical levels with the domestic/foreign distinction. The 'inside/outside' dividing-line is difficult to specify, and using state sovereignty as the break-point tells only half the story. The distinction itself, however, is still central to our understanding of world politics, not least because people perceive it to exist and to have a *constitutional* importance. Something called 'society' still exists despite the depredations of laissez-faire capitalism, and it survives in a great diversity of forms, usually within a cradle provided by the legally sovereign state. In consequence, despite the obvious challenges to their power and authority from transnational forces, governments have no option but to strive for at least a partial degree of control over the policy areas for which they are responsible. After all, their internal legitimacy, as Herz said, 'in our day is closely related to democracy in the broad sense of people having the conviction that they control their destinies and that government operates for their welfare.'[49] This means that the expectations for the elements of democracy and accountability in the world system largely fall on states. Where else can it reside? In UN-sponsored conferences on the environment? In the editorials of populist newspapers owned by multinational enterprises? In worldwide phone-ins to CNN? The history of political action shows that many claim to speak for the people, but few have the legitimate right to do so. Domestic society and its political institutions, imperfect as they are, represent the only way of expressing the authentic popular voice and assembling trade-offs in the complex process of coping with constraint and enlarging opportunity that international survival entails.

The key word here is 'responsibility'. We all feel responsibility for individuals we care about, and nationality will not be an issue, except in determining the direction of some of our attachments in the first place. But where does collective responsibility lie for the common weal, and who defines how far commonality extends in the international sphere? How do we decide who has prior claims, and how do we ensure the proper operation of 'exit, voice and loyalty', all of which imply the coexistence of separate, self-regulating communities?[50] It is to the credit of the political philosophers and international theorists that they have insisted on these normative questions over the last twenty years, but there has been far less discussion among the empiricists of the more practical aspect, namely, what is happening to responsibility and democracy under the impact of transnationalism, and how might the workings of the state be adjusted to cope with a transformed environment? Functionalism, postmodernism and globalisation theory have all in their different ways ducked this question, and foreign policy analysis has not been much better given its positivist bias.

The location of agency is one of the most fundamental questions of political science, and allowing that the answers vary, as Susan Strange argues, then there will still be identifiable internal and external dimensions to work with in every issue-area. It depends on what we are trying to explain. Why nuclear weapons have not proliferated more than they have requires a very different kind of account from the problem of why Germany and France differed over the recognition of Croatia, which will be different again from the problem of Nigeria's isolation and economic incompetence. The first demands a focus mostly on the international system, the second on domestic opinion and differing historical experiences, and the third on the interaction of the global oil market with a ramshackle state structure and deep ethnic divisions. All are important for different reasons, and in each it is impossible to say anything incisive without making use of some form of the domestic/foreign division.

What is more, in order to get a grip on the problem of how to act and for whom, we need to think in terms of 'the domestic sources of foreign policy' and its converse, 'the external sources of domestic policy'.[51] The first involves a very wide range of influences on a society's external projection: bureaucracy, constitution, parties, regions, culture, media. Some of these will have been significantly changed by transnational forces, the media prominent among them. But there will also still be a solid core of elements which appertain to a particular society, and are perceived to do so by outsiders. The generalisations about national character (read political culture) which spawn so many books are not without some basis, after all.[52] Conversely, there are clearly many identifiably external forces which impact upon a given social structure, and an efficient political process will sift them so as to decide which are healthy and part of the democratic flow of free exchange, and which are hostile or damaging in their consequences. Decisions will then have to be taken on what might be feasible to do about the latter, whether contaminated beef, Spanish fishing boats, human rights activists, espionage or the mere favouring of one electoral candidate over another. In Kahler's words, 'the state serves not only as a gatekeeper between domestic politics and the international environment; it can also serve to transmit international shocks into the domestic arena.'[53]

The shrinking of distance means now that no one can just 'mind their own business', and that isolation is not a realistic option even if a people might be willing to foreswear the benefits of economic growth.

But it is a far greater leap from this observation to the position that domestic society is losing its distinctiveness than some observers imagine. It still matters enormously as to the kinds of lives we live as to whether we are citizens of Pakistan or of Portugal. And if we do, on reflection, conclude that 'the domestic' still represents something meaningful to us, as both citizens and analysts, then it follows that a range of dealings with other people's domesticities will also be necessary. Those dealings which a government conducts on our behalf, collectively, we might as well call 'foreign policy'. 'Inside' does indeed imply 'outside', and vice versa. Wherever there is an existential 'us', decisions will have to be made about 'you', or 'not us'. These need not be hostile or ungenerous, but as the evolution of European Monetary Union is currently demonstrating, they will always be political.

To point these things out is not to fall back on state-centric realism, or to rehabilitate the anarchy problematic. It is, rather, an attempt to show that we will not do justice either to the differentiation of the human race or to our need for democratic political change if we carelessly neglect the problem of agency in international relations in favour of normative clarion-calls or systemic generalisation. The relationship between the domestic and the foreign is always in flux; but it is still crucial to our understanding of politics, that is, what has been done, what can be done, and what should be done.

NOTES

1. The structure-agency 'problem' has been well ventilated by Martin Hollis, Steve Smith and Walter Carlsnaes, after Alexander Wendt's initial formulations. See Alexander Wendt, 'The Agent-Structure Problem in International Relations Theory', *International Organization* (Vol. 41, No. 3, Summer 1987); Martin Hollis and Steve Smith, *Explaining and Understanding International Relations* (Oxford: Clarendon Press, 1990); Hollis and Smith, 'Two Stories about Structure and Agency', in *Review of International Studies* (Vol. 20, No. 3, July 1994); and Walter Carlsnaes, 'The Agency-Structure Problem in Foreign Policy Analysis', *International Studies Quarterly* (Vol. 36, No. 3, September 1992).
2. See the authoritative account of Andrew Linklater, 'Marxism' in Scott Burchill and Andrew Linklater (eds.), *Theories of International Relations* (London: Macmillan Press, 1996), pp.119–44.
3. Julian Saurin, for example, attacks 'the illegitimate methodological separation of "domestic" and "foreign", "internal" and "international".' Julian Saurin, 'The end of international relations?', in John Macmillan and Andrew Linklater (eds.), *Boundaries in Question: New Directions in International Relations* (London: Pinter, 1995), p. 258.
4. Chris Brown avoids most over-simplifications, but asserts that realism included the study of foreign policy decision-making, despite the fact that Foreign Policy Analysis grew up as a pluralist reaction to realism's blackboxing tendency. Chris Brown, *Understanding International Relations* (London: Macmillan Press, 1997), p. 67.
5. Cynthia Weber, *Simulating Sovereignty: Intervention, the State and Symbolic Exchange* (Cambridge: Cambridge University Press, 1995), pp. 10 and 6. See also pp. 24–5.
6. R.B.J. Walker, *Inside/Outside: International Relations as Political Theory* (Cambridge: Cambridge University Press, 1993), p. 25.
7. *Ibid.*, p. 171.
8. R.B.J. Walker, 'On pedagogical responsibility: a response to Roy Jones', *Review of International Studies* (Vol. 20, No. 3, July 1994), p. 316. Walker was replying to Roy Jones' criticisms of *Inside/Outside* in the same issue, entitled 'The Responsibility to Educate'. This exchange is illuminating and enjoyable. Unlike N.J. Rengger, however, I do not conclude that Walker managed to out-point Jones; N.J. Rengger, 'Clio's cave: historical materialism and the claims of "substantive social theory" in world politics', *Review of International Studies* (Vol. 22, No. 2, April 1996), p. 219.
9. Ken Booth, 'Human Wrongs and International Relations', (text of the second John Vincent Memorial Lecture) *International Affairs* (Vol. 71, No. 1, January 1995).
10. *Ibid.*, p. 110.

11. *Ibid.*, pp. 120–21.
12. Booth is careful not to use the virtually discredited term 'imperialism'. But there is no doubt that he believes the state system does represent the imposition of dominance by the rich over the poor (*ibid.*, p. 121). See also Ken Booth, '75 years on: rewriting the subject's past – reinventing its future', in Steve Smith, Ken Booth and Marysia Zalewski (eds.), *International Theory: Positivism and Beyond* (Cambridge: Cambridge University Press, 1996), pp. 328–339 (especially p. 336).
13. Jan Aart Scholte, *International Relations of Social Change* (Milton Keynes: Open University Press, 1993).
14. This loose term subsumes at one and the same time liberal beliefs in the growing importance of regimes and institutions, free-marketeers' trust in the historical destiny of capitalist enterprise, and the critical theorist's perception of a 'global hegemony' operating 'through alliances between elites in core and industrializing societies'. Linklater, *op.cit.*, p.133; see also Mark Webber, 'States and statehood' in Brian White, Richard Little and Michael Smith (eds.), *Issues in World Politics* (London: Macmillan Press, 1997), pp. 41–42; and A.J.R. Groom and Dominic Powell, 'From world politics to global governance' in A.J.R. Groom and Margot Light, (eds.), *Contemporary International Relations: A Guide to Theory* (London: Pinter, 1994), pp. 81–90.
15. A problem discussed in Andrew Marr, *Ruling Britannia: the Failure and Future of British Democracy* (London: Penguin, 1996), pp. 187–209. The key issue, of course, is not bucking or accepting the world economy *in toto*, but rather how to handle global trends in differentiated ways that serve national and local needs.
16. Robert Keohane and Joseph Nye (eds.), *Transnational Relations in World Politics* (Cambridge, MA: Harvard University Press, 1970); Keohane and Nye, *Power and Interdependence: World Politics in Transition* (Boston: Little, Brown, 1977).
17. Susan Strange, *States and Markets: An Introduction to International Political Economy* (London: Pinter, 1988). In fact by this title Strange did not mean to imply that states were the only source of authority; she also includes churches, guilds, cartels, media, banks etc.
18. Steve Smith, 'Power and Truth: A Reply to William Wallace', *Review of International Studies* (Vol. 23, No. 4, October 1997), pp. 507–16. Smith argues: 'What politics seems to me to be crucially about is how and why some issues are made intelligible as political problems and how others are hidden below the surface (being defined as "economic" or "cultural" or "private").' Smith is not so obviously an ethical globalist, but his view of epistemology carries him in that direction.
19. See Chapter 11, 'Ecology as Religion' in Matt Ridley, *The Origins of Virtue* (London: Viking, 1996), pp. 211–25.

20. See John Vogler, 'Environment and Natural Resources', in White, Little and Smith, op.cit., pp. 222–45.
21. *Op.cit.*, p. 41.
22. *Ibid.* p. 37.
23. The extensive debate which has raged since Michael Doyle's article of 1983, on whether democracies are more or less war-like than other kinds of state needs to be linked to discussions of foreign policy and political responsibility. A good recent overview can be found in Nils Petter Gleditsch and Thomas Risse-Kappen (eds.), *Democracy and Peace*, special issue of the *European Journal Of International Relations* (Vol. 1, No. 4, December 1995).
24. Both 'contractual' and 'affective' in this context are forms of communitarianism, the antagonist of the cosmopolitanism discussed above. The communitarian versus cosmopolitan debate is a useful ground-clearing exercise. But the dilemma for most of us in the late twentieth century is not whether our loyalties lie more with our group or with humanity as a whole, but how – morally and practically – to *reconcile* the various obligations we feel, to our ailing neighbour and to the victim of torture in a foreign gaol, to our unemployed co-nationals and to the exploited workers of a Third World sweatshop. Chris Brown shows that this has always been a central dilemma in international theory. Chris Brown, *International Relations Theory: New Normative Approaches* (London: Harvester Wheatsheaf, 1992), for example, p. 111.
25. Television has introduced a twist in recent years, with millions of people feeling as close to the characters of foreign soap operas as to their own local heroes. One of the most popular has been the Australian show ironically called 'Neighbours', but Russians have become addicted to the doings of Mexican housewives and Egyptians have enjoyed *Dynasty*.
26. Terms easily confused with the 'society' or 'community' of states, discussed in Hedley Bull, *The Anarchical Society* (London: Macmillan Press, 1977) and James Mayall (ed.), *The Community of States* (London: Allen & Unwin, 1982). See in particular Christopher Brewin on Collingwood's view of community and society, pp. 45–47. I take 'community' to be prior in the sense that it involves an important shared sense of identity, and 'society' to be prior in that it involves shared rule, or *Recht*. To function well a nation-state needs both.
27. See Robert D. Putnam, with the assistance of Robert Leonardi and Raffaella Nanetti, *Making Democracy Work: Civic Traditions in Modern Italy* (Princeton, N.J.: Princeton University Press, 1993).
28. 'Mixed actor' is the term used by Brian Hocking and Michael Smith in their *World Politics: An Introduction to International Relations* (London: Harvester Wheatsheaf 1990), pp. 178–79.

29. The origins of the word lie in the old French 'forain', and the Latin 'foris', meaning 'outside'. *The Oxford English Dictionary*, Second Edition, Vol.VI (Oxford: Clarendon Press, 1989), p. 51.
30. On the problems involved in defining foreign policy, see Christopher Hill, 'Foreign Policy', in Joel Krieger (ed.), *Oxford Companion to World Politics* (New York, NY: Oxford University Press, 1993).
31. This scholarship derives from both history and political science. Among the leading examples are Robert Louis Gallucci, *Neither Peace Nor Honour: The Politics of American Military Policy in Viet-Nam* (Baltimore, MD: Johns Hopkins University Press, 1975); Guenter Lewy, *America in Vietnam* (New York, NY: Oxford University Press, 1978); Leslie H. Gelb and Richard K. Betts, *The Irony of Vietnam: the System Worked* (Washington, D.C.: the Brookings Institution, 1979); and George C. Herring, *LBJ and Vietnam: A Different Kind of War* (Austin, TX: University of Texas Press at Austin, 1994). For a wider sample, see Robert J. McMahon (ed.), *Major Problems in the History of the Vietnam War: Documents and Essays* (Lexington, MA: D.C. Heath, 1990).
32. See Ralph B. Levering, *The Public and American Foreign Policy, 1918–1978*, (New York, NY: Morrow, 1978), and Ole R. Holsti, *American Opinion and American Foreign Policy* (Ann Arbor, MI: University of Michigan Press, 1996).
33. This perception produced one of the finest of all examples of war literature, Michael Herr, *Dispatches* (New York, NY: Alfred Knopf, 1977).
34. Cited in Robert A. Divine, 'The Johnson literature' in Robert A. Divine (ed.), *The Johnson Years, Volume One: Foreign Policy, the Great Society and the White House* (Kansas, KS: University Press of Kansas, 1987), p. 14. There are many ironies here, as some black fellow-citizens were quick to point out: 'Inasmuch as you have seen fit to send observers to Vietnam to see that "free and democratic" elections are held... it would mean much more to America and particularly 22 million Negroes if you would use your influence and call for new elections in many sections of Mississippi and send representatives to make certain Negroes and Negro candidates are assured justice and fair play in all elections.' Telegram from Charles Evers and Aaron Henry to Johnson, 30 August 1967, Johnson Library. Cited in Steven F. Lawson, 'Civil Rights', in Divine, p. 122.
35. Susan Strange, 'Territory, state, authority and economy: A new realist ontology of global political economy', in Robert Cox (ed.), *The New Realism* (Tokyo: The United Nations University, 1992).
36. Discussed in Christopher Hill, 'Implications of the World Society Perspective for National Foreign Policies', in Michael Banks (ed.), *Conflict in World Society: A New Perspective in International Relations* (Brighton: Harvester/Wheatsheaf, 1984).

37. A theme identified by James Mayall, in 'The Variety of States' in Cornelia Navari (ed.), *The Condition of States* (Milton Keynes: Open University Press, 1991); see also Robert H. Jackson, *Quasi-states: Sovereignty, International Relations and the Third World* (Cambridge: Cambridge University Press, 1990).
38. This point was made vigorously by F.S. Northedge in 'Transnationalism: the American Illusion,' *Millennium: Journal of International Studies* (Vol. 5, No. 1, Spring 1976), pp. 21–7.
39. Geoff Mulgan (Head of the Demos think tank used by Tony Blair), 'Myth of Withering Government', *The Independent*, 15 May 1995.
40. The exception is the United States, of course. In competing with the US, the EU has to cope with its own Euro-domestic factor of immobile labour, for historical and linguistic reasons.
41. A conclusion broadly endorsed by Ian Clark, who argues that there is likely to be 'a new accommodation between state power and the forces of globalization, rather than the outright victory of the one over the other.' Ian Clark, *Globalization and Fragmentation: International Relations in the Twentieth Century* (London: Oxford University Press, 1997), pp. 195–96.
42. This notion, now once more receiving attention after long neglect, is explored in Christopher Hill, 'World Opinion and the Empire of Circumstance', *International Affairs* (Vol. 72, No. 1, January 1996). See also Frank Louis Rusciano and Roberta Fiske-Rusciano, 'Towards a Notion of World Opinion', *International Journal of Public Opinion Research* (Vol. 2, No. 4, 1990); Frank Louis Rusciano, 'Media Observations on World Opinion during the Kuwaiti Crisis: Political Communication and the Emerging International Order', *Southeastern Political Review* (Vol. 24, No. 3, September 1996); Frank Rusciano, Roberta Fiske-Rusciano and Minmin Wang, 'The Impact of "World Opinion" on National Identity', *Harvard International Journal of Press/Politics* (Vol. 2, No. 3, 1997), pp. 71–92.
43. After Rabin was killed in 1995, Shlomo Avineri observed that the 'most poignant moment' at his funeral had been when King Hussein of Jordan had spoken of the assassination of his own grandfather King Abdullah in 1951. Abdullah had been killed by a Palestinian for negotiating peace with Israel; Rabin was murdered by a Jew for negotiating peace with the Palestinians. Shlomo Avineri, 'Murder in Zion', *The World Today* (Vol. 51, No. 12, December 1995), p. 226.
44. '"Donne del mondo: aiutate le afghane". Appello della Bonino all'Occidente', *La Repubblica*, 7 October 1996. The Talibans have so far proved unmoved.
45. The draconian penalties for drugs smugglers are a case in point. But it should be remembered that this is very much a top-down morality, directed against human rights universalism as much as hippy individualism. See Yash Gai, 'Asian Perspectives on Human Rights', in James T.H. Tang (ed.),

Human Rights and International Relations in the Asia Pacific (London: Pinter, 1995), pp. 54–67.
46. Friedrich Kratochwil, in this volume, p. 203.
47. See Brian Hocking (ed.), *Foreign Relations and Federal States* (London: Leicester University Press, 1993).
48. Both Mexico and Canada would see a political NAFTA as formalising US dominance in North America. It is early days yet for Mercosur. See *The Economist*, Special Survey on Mercosur, 12–18 October 1996.
49. John Herz, 'The territorial state revisited: reflections on the future of the nation-state' in James Rosenau (ed.), *International Politics and Foreign Policy* (New York, NY: The Free Press, 1969), p. 83.
50. The terms are from the famous book by Albert O. Hirschman, *Exit, Voice and Loyalty: Responses to Decline in Firms, Organizations and States* (Cambridge, MA: Harvard University Press, 1970), in which he distinguished between the three strategies open to a citizen: opting out (possibly including migration); political action; and acceptance or acquiescence, usually predicated on trust in the likelihood of things improving. It is this last strategy – loyalty – which is particularly interesting for our purposes. Hirschman shows that far from being inherently irrational, loyalty is often highly functional.
51. James Rosenau was, I think, the first to use the term 'the domestic sources of foreign policy' in his book of the same title (see his pioneering chapter on 'Foreign policy as an issue-area'), although Richard C. Snyder talked of the 'internal setting' of foreign policy back in 1954. James Rosenau, *The Domestic Sources of Foreign Policy* (New York: Free Press, 1967). Peter Gourevitch drew attention to the converse in his 'The Second Image Reversed: the International sources of Domestic Politics', *International Organization* (Vol. 36, No. 4, 1978), pp 881–912. For a perceptive use of the latter approach, see Miles Kahler, *Decolonization in Britain and France: The Domestic Consequences of International Relations* (Princeton, NJ: Princeton University Press, 1984). Kahler shows how in some respects empires can colonise metropoles.
52. For a subtle and entertaining commentary on the dialectics of difference and similarity across national boundaries see Hans Magnus Enzenberger, *Europe, Europe* (London: Hutchinson, 1989).
53. Kahler, *op.cit.*, p. 379.

8. The Politics of Place and Origin: An Inquiry into the Changing Boundaries of Representation, Citizenship, and Legitimacy
Friedrich V. Kratochwil

There seems to be a growing disparity between the practices which comprise international relations, and the conceptual apparatus by which we attempt to analyse these practices. This divergence is most obvious in the case of sudden fundamental change, as in the case of the disappearance of the Soviet Union, or the (re)emergence of the problematique of nationalism. Such cases provide crucial evidence which contravenes the adequacy of our conceptual apparatus and the theories based on it. Nevertheless, denials are more common than serious reconceptualisation. Since Yosef Lapid and I have dealt with this phenomenon, exemplified by the neo-realist treatment of nationalism in a separate paper[1], I do not want to rehearse those arguments here any further.

Equally surprising, though, is the fact that the treatment of long-term secular changes seems to have little impact on the conceptual lenses by which we analyse the ongoing practices of international politics. There is, for example, a growing literature on the inadequacy of our traditional understanding of the state as a unit, based on the impact of growing interdependence, the communications revolution, and the globalisation of production, that undermines the capacity of states to make autonomous choices. But if autonomy represented a significant part of our understanding of sovereignty, then the basic building block on which our understanding of international politics was based becomes increasingly problematic. The 'internationalisation of the state', i.e. international regimes, on the one hand enables states to pursue their interests, but also restricts the domain of their autonomy. Similarly, the conjunction of international migration and human rights, or the development of a global civil society, all suggest that new conceptual tools have to be forged in order to understand politics at the end of the century. David Held has discussed this phenomenon in terms of the several *disjunctures* that political theorists, international relations scholars and students of the state alike encounter. The 'disjunctures'

> reveal a set of forces which combine to restrict the freedom of action of governments and states by blurring the boundaries of domestic politics, transforming the conditions of political decision making, changing the institutional and organizational context of national politics, altering the legal framework and administrative practices of governments, and obscuring the lines of responsibility and accountability of national states themselves. These processes alone warrant the statement that the operation of states in an even more complex international system both limits their autonomy and impinges increasingly upon their sovereignty. Any conception of sovereignty which interprets it as an illiminitable and indivisible form of public power is undermined. Sovereignty itself has to be conceived today as already divided among a number of agencies – national, regional and international – and limited by the very nature of this plurality. [2]

The upshot of the above argument is that an analysis that examines the concepts singly and in term of simple dichotomies – the favourite pastime of empiricists that either something is, or is not the case – is seriously misleading, since it neglects to inquire into the changing meaning of related concepts within an entire semantic field.

It is the task of this chapter to examine the changing relations among the concepts of sovereignty, citizenship, and legitimacy. I argue that by investigating these three fundamental concepts and their relations we gain crucial insights into the changing meaning of politics at the century's end. Such an enquiry seems mandated not by purely academic interest, but by the increasing divergence between the *demos*, on the one hand, and the 'subjects' who, according to democratic theory, are understood to be the 'citizens' who legitimate decisions by their 'consent'. Liberal theory in particular has focused on the issue of 'consent' as a validator of decisions, given inevitable hold-out problems and the emergence of strategic behaviour among the participants in a political setting. The relevance of such an inquiry is further enhanced by a second problem. At present, massive international migrations are calling into question the very notion of a national community of fate that was fundamental to the modern sovereign democratic state.

While I am certainly not able to solve the practical problems that arise in this context of deciding who belongs to the 'us', I still hope to clarify some of the issues. In particular, I think that the dynamics of politics emerges from the definition of the group, the efforts of locating it in a place and in establishing representative institutions entitled to make public choices. To that extent, membership (citizenship), sovereignty and legitimacy cover this *problematique*.

In elaborating on these issues, my argument takes the following steps. In the next section I begin with the last problem mentioned, i.e. legitimacy and try to show that 'consent' as the legitimising source of political authority fails and necessitates a theory of representation. Unfortunately, and contrary to the liberal claim that 'the people' are the ones represented, it can be shown that the language game of representation comprises considerably more complex issues, quite aside from begging the question of who 'the people' are, that can advance claims to representation. In section three, I analyse two conceptions of 'the people', one more influenced by classical republican and participatory images, and one more by cultural and historical myths of a common descent. I also want to show how the state, defining itself both as a membership organisation and as a territorial/jurisdictional sphere, has always served as a mediating organ by ascribing the status of 'citizenship'. This ascription was either skewed towards the 'political' or towards the 'cultural' side, but could never be decided according to a 'pure' model of political participation or of origin. These considerations in turn open the conceptual space for the discussion of the dimensions and problems of citizenship in the fourth section. A brief summary in section five concludes the chapter.

THE PROBLEM OF CONSENT

One of the fundamental tenets of liberal theory is that the exercise of public authority has to be based on the consent of the governed. This principle has two corollaries: one, that the actions of the public authority are legitimate only insofar as they can be construed as the result of the consent of the subjected, and two, that there should be a congruence between 'the people', and the jurisdiction of the public authority. It is the latter problem that leads in case of large scale migrations to a certain legitimation crisis of modern states, although historically, the existence of large segments of 'foreigners' (*metoikoi*) was not unusual. Not only is the symmetry between the governing and the governed disturbed by the existence of such 'aliens', but the legitimacy of governmental acts *vis-à-vis* foreigners has to be inferred somehow from some kind of act, such as entering the country or transacting business there.

Although this is the standard account since Locke made the argument that even travelling on the highways of a foreign country implies 'consent', (as does the owning of property), such an argument is open to a variety of objections. First, if one is to infer consent from some silent or non-explicit act, then such an inference seems valid only if there was an antecedent rule that such and such an act meant precisely the uptake of an obligation. In the absence of an institutional rule to that effect, it is hard to fathom how one could infer normative consequences from merely factual observations. This leads to a second difficulty: if this rule is construed as the crossing of the boundary by a 'foreign' individual against whom the state asserts jurisdiction, it is the state's assertion of jurisdiction which does most of the explaining and the person's 'consent' hardly does any work. As Lea Brilmayer aptly put it:

> The implicit assumption amounts to a prior assumption of state territorial sovereignty. Only a state that has territorial sovereignty may condition entrance upon consent to obey the law. If the state possesses territorial sovereignty, however, reliance on defendant's consent, whether explicit or implicit is unnecessary. Consent is largely superfluous; indeed it only serves to mask the fact that territorial sovereignty provides the real basis for the exercise of personal jurisdiction.[3]

Here, a *third* problem emerges that makes the individualistic consent metaphor problematic, unless it is buttressed by some notion of representation. Brilmayer's remark that only states can do certain things, reminds us that acts of sovereignty are *sui generis*, i.e. are of a different quality and can therefore not easily be interpreted as the result of individual delegation, even if several of these individual and private delegations are summed up. The coming into existence of a corporate body creates a new space among the individuals so constituted, and the new rights and duties belonging to this corporate body cannot be said to consist of individual rights that have been delegated. Corporate law speaks in this context of the 'corporate veil' that cannot be lifted unless, and until, the corporate body no longer exists.[4] It seems to suggest that at least one of the reasons Rousseau tried to distinguish the *volonté des tous* from the *volonté général,* was perhaps to call attention to this new quality that emerges through the constitution of a number of people as a body politic. In a similar vein, the sovereign becomes, in Hobbes's construct, the 'fixer of signs', part of which is the competency to decide on its own competency. It is, therefore, difficult to derive such a power from the notion of the sovereign as an arbiter of and enforcer of delegated private rights.

The peculiarity of this incorporation, and the emergence of the public space is further shown by the intended duration and practical irrevocability of the contract. As Locke put it:

> he that has once, by any express declaration, given his consent to be of any commonwealth, is perpetually and indispensably obliged to be and remain unalterably subject to it, and can never be again in the liberty of the state of nature; unless by any calamity, the government he was under, comes to be dissolved; or else by some public act cuts him off from being any longer a member of it.[5]

This is all the more important as Locke distinguishes clearly between the implied and explicit consent binding the members of a commonwealth. Implied consent is indicated by the enjoyment of property. But after sale or donation of such possession, this consent dissolves and leaves the individual free to 'to go, and incorporate himself into any other Commonwealth, or to agree with others to begin a new one, in *vacuis locis,* in any part of the world, they can find free and unpossessed.'[6] Thus, while it is only express consent 'which makes any one a member of any Commonwealth'[7] it is not quite clear within the contractual paradigm why the commitment shall be insoluble, and also not be terminable by an *actus contrarius* if not by the very act of alienating one's property, as Locke himself seems to suggest.

A *fourth* difficulty with a strict consent theory and of its attendant political obligation (both of which are the two sides of the same argument justifying governments and their actions) emerges when we consider the problem of hold-outs which, in private law, simply result in the abortion of transactions, a result which, in the case of a body politic, however, could be fatal. It is here that Locke introduces as a remedy the notion of a 'majority' that can oblige even recalcitrant or opposing members.[8] But in that case the notion of 'consent' becomes strained as Locke himself realises. While he suggests that the construction of an act backed by a majority as an act of the whole is problematic since 'nothing but the consent of every individual can make anything to be the act of the whole,' he also admits that 'such a consent is next to impossible even to be had.'[9] This ambivalence of relying on principles other than consent while wanting to make it appear that 'consent' still serves as the foundation goes through the entire *Second Treatise*. Thus Locke declares on the one hand that 'the supreme power cannot take from any man any part of his property without his own consent', but he also states, when faced with the hold-out problem and the need to procure public goods by means of taxation, that 'a government cannot be supported without great charge [and] 'tis fit, everyone who enjoys his share of protection, should pay out of his estate his portion for the maintenance of it.' But still it must be with his own consent, i.e. the consent of the majority, giving it either by themselves, or their representatives chosen by them.[10]

'Never has an innocent Latin abbreviation (i.e.),' Don Herzog aptly remarks, 'done more work in an argument.'[11] Locke not only magically transforms the necessary individual consent into one of *the majority*, but he also makes it into one of a *majority in parliament*. Individual consent is, thereby, twice removed as a legitimising force. It also happens that without the silent, or even explicit reliance on other principles (such as functional necessity) or on practices that have attained some stability and standing, the question of legitimacy would be indeterminate. It has been the merit of Herzog's analysis to have uncovered the political roots of the particular arguments that owe more to the events and political practices of the English Revolution and the Stuart Restoration than to their conceptual coherence. Herzog notes:

> Against the practices of loyalty oaths and the virtual blackmail by the King as well as by Cromwell – who extracted 'loans' and 'benevolent' contributions from their subjects – parliament as the assembly of power holders had means of refusing such 'hold ups'

which individuals did not. Parliamentary consent was preferable not because August representatives make better choices, or because Parliament was somehow a perfect microcosm of all England', a rather problematic construction given the extremely limited franchise and, thus, the 'unrepresentative character of this assembly. But parliament as a 'corporate body had powers that individuals didn't. Locke's hastily changing the individual's own consent to taxation to the consent of a parliamentary majority is then not a gaffe... To the contrary, it's perfectly sensible in light of these practices.[12]

These critical remarks have driven home the fact that, given the special character of the political order, consent theorists have to fall back on other principles in order to deal with these special problems. Two were of particular importance in this context: representation and membership. While the issue of representation has received most of the attention in subsequent treatises and has become one of the mainstays of democratic political theory, the membership question has hardly ever been subjected to scrutiny. This is all the more surprising as the clarification of who 'the people' are is obviously of great importance, especially if a majority vote is introduced as one of the reasonable means of dealing with hold-out problems. Locke himself seems to have realised, at least indirectly, the importance of this question by making membership irrevocable. A stable body politic can only be formed if entry and exit options are constrained. Membership in a political community apparently cannot be fashioned analogous to the contractual paradigm of simple exchanges.[13]

Nevertheless, the predominant tendency of liberal theory has been to assume that this membership in particular political societies is either somehow illegitimate, or that the established communities provide the unproblematic units of analysis. The protagonists of these two versions are Locke on the one hand, and Rawls on the other.

For Locke it is clearly mankind as a whole that represents the proper subject for the inquiry into the foundations of political order. To that extent, the facts of historical singularity are of no significance since 'were it not for the corruption and viciousness of degenerate Men, there would be no need... that men should separate from this great and natural Community (Mankind), and by positive agreements combine into smaller and divided associations.'[14] For Rawls, on the other hand, the theory of justice is no longer a universal but a political project,[15] i.e. an attempt to specify the principles of justice for societies that have been historically constituted. But despite this, Rawls's argument concerning the choice behind the veil of ignorance, his complete rejection of any historically grounded principles of dessert, his insistence on the irrelevance of practical experience with these principles and their possible revisions in the light of such lessons, all suggest that he is after something far more ambitious and demanding.

The internal contradiction, at least in the early Rawls, is indicative of the tension between, on the one hand, the universalism of classical liberalism as a foundationalist account of political order – often indebted to some intuitionist version of natural law – and, on the other hand, the recognition of historicity. The latter can no longer invoke some intuitive natural law and has to accept the multiplicity of political societies. Nevertheless, an analytical philosophy *á la* Rawls must consider the historically constituted separateness of the political units not a matter of significant moral import, otherwise the principles of justice lose their obligatory force. But as Hume suggested in his 'anti-foundationalist' *History of England*,[16] it might be due to these very historical peculiarities and settled habits, rather than to the coherence of political theories or constitutional principles, that political systems function. Similarly, Weber's argument[17] regarding legitimacy, when properly cleansed of the rather debatable implications of a universal and unidirectional historical trend towards modernity, provides considerable food for thought. Indeed, traditional forms of legitimacy share certain common myths embedded in narratives and collective memories. But the claim to rule is always the result of *particular* events and of their later reconstructions in the light of specific political problems a community faces. To that extent, political scientists had better examine these peculiarities instead of treating them merely as contingencies which can be neglected.

This preliminary inquiry into the logic of consent theory as an explanatory as well as a justificatory device has done nothing to encourage optimism in formulating clear universalist criteria. Since issues of representation and membership loom large on the horizon, it seems clear which direction such a theoretical revision has to take. It is the task of the next section to examine more closely the two concepts of representation and membership and their function in the larger semantic field of legitimacy.

THE PUZZLES OF REPRESENTATION AND MEMBERSHIP

To represent someone or something is, according to the etymology of the term 'to make present', something or someone who is not there. To that extent, the issue appears to be straightforward. Representing 'the people' in a political sense seems to provide something akin to a map that informs us about the attitudes, desires, or interests of the population at large, in order to make public choices. Two questions arise in this context: first, representation by resemblance attempts to provide information, but not all information is relevant, as even the most accurate maps cannot show everything. The key problem is which characteristics are politically relevant and worthy of reproduction. The 'history of representative government and the expansion of the suffrage is one long record of changing demands for representation based on changing concepts of what are the politically relevant features to be represented.'[18] Property requirements or restrictions of political rights to free males belong here, as do the debates about various electoral systems and their respective distortions. But most important, of course, is the question which remains in the background, i.e., who 'the people' are, in other words, the universe which requires representation and makes itself 'present'.

Second, at least in liberal theory, there are not simply amorphous people but specific *interests* of these people that are to be represented. But how does their representation help in making legitimate public choices? After all, we know from public choice that no common preference ordering can be established by simple methods of aggregation from three or more non-identical individual preference orderings. Here, James Madison provides an interesting answer. Not simply relying on exhortations of public spiritedness, or on the *volonté général*, he emphasises the ubiquity of factions generated by diverging interests. But he also sees that institutions can play a decisive role by selecting the relevant features of the 'will of the people'. In this context his argument about the advantages of a republican form of government become crucial. It is this form which is supposed to cure the ills of direct democracy.

Through representation, the business of government is entrusted in a *republic* to a few men rather than to the people at large, and the founding father hopes that through this representation the passions of factionalism will be abated. In other words, representation serves like a filter to

> refine and enlarge the public views, by passing them through the medium of a chosen body of citizens, whose wisdom may best discern the true interests of their country, and whose patriotism and love of justice will be least likely to sacrifice it to temporary or partial considerations.[19]

Like Hobbes, who suggested that through the establishment of a sovereign the expectations of individuals radically change, and that, therefore, the laws of nature are no longer obligatory only *in foro interno*, so Madison hopes to produce a refinement of interests by institutional means. Even his second and most famous suggestion, i.e. that through the expansion of space interests would become more varied and prevent the domination of one group, has a certain institutional ring. Here the danger of tyranny by the majority is abated by the existence of distinct states. 'The influence of factious leaders', he writes, 'may kindle a flame within their particular States, but will be unable to spread a general conflagration through the other States.'[20]

With this argument concerning the blessings of the republican form of government, we reach a rather different understanding of representation which transcends or competes with the descriptive meaning of the term. How such an institutional version of representation can mediate the notions of 'the people' or their 'interests' can easily be seen when we call judges, for example, 'representatives' of a political system. It is not 'interests', but the notion of the public order itself, whose agent the judge is, that now comes into focus. In this way the judge becomes a representative in two ways: as an elected official, as is done in some countries, or as a representative, standing for the public order symbolically. However, we have to keep in mind that although being a judge involves some symbolic representation – signalled by the special settings, robes, and formulaic invocations – a judge is not *merely* a symbol as the following considerations suggest.

First of all, it is important to realise that symbols 'represent' by not conveying particular information about the matters for which they stand. To that extent, symbolic representation is radically different from the descriptive representation with which we began our discussion. Indeed, we do not perceive or evaluate symbols by their descriptive accuracy, as there are no correct or incorrect symbols. Rather, what is at stake in symbolic representation is that the symbol evokes certain feelings and emotions,[21] and, thus, becomes the 'focus of attitudes thought appropriate.'[22] Second, purely symbolic representation, as, for example, the flag representing the United States, or even the Queen representing England, does not depend on actions for its effectiveness as symbols. Symbols represent largely by their *being,* and not by their ability to make choices on behalf of the people. Indeed, the representative function of a king in a democracy consists precisely in restricting his public choices made on behalf of his subjects.

The similarities and differences of the symbolic representation of a judge now become clearer. When a judge pronounces judgments in the 'name of the people' she is a representative of the public order by deciding a case and controversy in accordance with the rules that establish the limits of her discretion. To that extent, she is the truer representative the less she is swayed by public opinion or by the organised interests of the society. Indeed, legal theorists such as Ronald Dworkin have taken great pains to demonstrate the strict rules to which judges have to submit in order to arrive at decisions and thus to become representatives of the constitutional order.[23] Similarly, the limits imposed on the law-making capacity of representative institutions, such as the First Amendment which enjoins the U.S. Congress from restricting the freedom of speech, as well as the theoretical possibility of an unconstitutional amendment discussed among American legal scholars,[24] demonstrate that we use the term 'representation' and 'representing' in contexts quite different from the first descriptive and iconic account above.

Our puzzlement concerning the true meaning of representation is still further enhanced when we consider two more cases which are also part of the grammar of representation. Consider in this context the two versions of formalistic theories of representation, one focusing on the initial act of authorisation, the other looking at the problem of accountability for actions performed. Both are formal, as they concern only the conditions which determine the capacity for an agent to act on behalf of others. While Hobbes is particularly interested in the act of authorisation, accountability theorists stress the importance of holding the representative accountable after she has acted on behalf of the community. In the latter case, it is not the right of acting on behalf of others which is at issue, but rather the control of the decisions made by the representative.

It is not difficult to see that Hobbes's account of representation is built on the example of authorisation. Fundamental to Hobbes's paradigm is the distinction between an actor, i.e. the person who acts, and the author, which provides the actor with the right to act. Representation concerns virtually exclusively the issue of how a multitude authorises an agent to act on its behalf.

A multitude of men, are made *one* person, when they are by one man, or one person represented; so that it be done with the consent of everyone of the multitude in particular. For it is the *unity* of the representer, not the *unity* of the represented that maketh the person *one*. And it is the representer that beareth the person, and but one person; and unity, cannot otherwise be understood in multitude.[25]

Two things are important here; first, the argument that the unity of the people is constituted by the representer, not by some communality of all the people. Second, the notion which attributes the actions of the sovereign to everyone. In this sense Hobbes reminds us that someone punished by the sovereign is 'the author of his own punishment.'[26] Thus both notions together explain why the sovereign possesses powers and rights over whose exercise the subjects have no control, while these actions can still be ascribed to the individual subjects. Within the bounds of the newly constituted authority, the sovereign alone can judge how to make use of them, as there is no duty owed to the subjects to consult them, or to act according to their instructions or their interests.

It is precisely this latter point which *accountability theorists* of representation stress. For democratic theory based on this idiom of accountability the decisive institutional feature becomes elections. They are not primarily seen as mechanisms for the selection of representatives who are then authorised to make policy, but as an institutional measure for holding the representatives accountable for their actions. Actually, accountability might also be ensured by other means, as e.g. the appointment of representatives on administrative boards is designed to ensure responsiveness of the administrative officials. Here, the transition of the accountability theory of representation to the descriptive theory of representation becomes visible, even though the emphasis is more on ensuring the responsible exercise of power, and less on the representation of 'the community' or 'the people'.

Our analysis seems to turn on itself. Puzzled by the problem of consent which relied on representation for much of its persuasive strength, we encounter a new puzzle: there is no clear concept of representation. Instead, a variety of closely related though quite distinctive practices and conceptions of making public choices are covered by this term. Even more importantly, it seems that we cannot decide which of the different cases of representation is the accurate one, since all of them seem to exemplify at least one genuine aspect of representation. Each of the different models seems to be derived from one example which quickly loses its plausibility when it is applied to other contexts that are illuminated by a different example. Representation is thus more like a cluster-term for a variety of understandings that might share a certain family resemblance, but that do not seem to have a firm core aside from the fact that it serves as a cipher pointing to the importance of 'the people'.

It is, therefore, to this problem and to the practices of inclusion and exclusion that we have to turn our attention. Making membership and the dichotomy between friend and foe the fundamental categories for 'the political', is not only characteristic of Carl Schmitt; it has a long tradition dating back to Aristotle and Plato who focused on participation, ancestry, and the land as defining criteria of the political. Thus choosing membership as a critical category for this inquiry does not entail sharing Schmitt's concerns or criticism of Western democracies. It follows rather cogently from the need to understand the practices by which we decide whether one is 'inside' or 'outside', whether one is a 'member' or an 'other'. This does not mean that one has to be a foe; here Schmitt's dichotomy appears hasty and unwarranted. True, one *can* be the 'foe', the 'foreigner', with whom one has little in common even if relations are peaceful and without animosity. But one can also be admitted to sharing a place with a group, but not to being a member of it, even if, or particularly when the group understands itself not simply as a multitude, or as a fortuitous aggregation, but as an ongoing transgenerational enterprise, as a 'community of fate'.

This community has to be located somewhere, it has to have a place in the world. Thus it is not surprising that the *land* often defines the group or indicates the origin from where the group came or is invoked in order to evoke a sense of permanence and identity. And it is out of this dynamic between defining the group and locating it in certain places that the drama of politics emerges.[27] After all, the term politics derives from *polis*, which means to build a wall. These walls include and exclude members as well as delineate the space that is home, by setting it apart from the wilderness, the no man's land, or from the land of others. Locke's argument, quoted above, comes to mind. He speculated about the legitimacy of leaving the political association, provided one had not been part of the social contract, could find others with whom to contract, and also could find *unoccupied land* in order to found a new civil society and then a government.

But if states are membership as well as territorial organisations, and if they are transgenerational rather than purely contractarian enterprises, then the drawing of lines has less to do with voluntary undertakings and individual choices, and more with the ascription of *status*. 'To be defined as a citizen', writes Rogers Brubaker, 'is not to qualify as an insider for a particular instance or type of interaction; it is to be defined in a general, abstract, enduring, and context-independent way as a member of that state.'[28]

This status is difficult to shun. It imposes special duties that are not part of one's own choosing, and gives members certain rights that others may not claim. This status is usually ascribed at birth, because growing into the role will shape the attachments, life plans, interests, and, thereby, provide the emotional bonds and loyalties that transcend the immediate family by also tying us to our fellow-citizens. This does not mean that citizenship is necessarily primordial,[29] so fundamental that it cannot be changed, but it *does* mean that such changes are not easy, for individual as well as for political reasons. True, our *Wahlverwandtschaften* (chosen attachments) can, as Goethe reminds us, be as close and important as the primordial ones. But changing membership status necessitates the individual to not only sacrifice earlier loyalties, but such an act usually also implies that at least parts of personal identity have to be reshaped. On the political level it means that the applicant has to navigate carefully among the numerous formally articulated rules which states promulgate in order to attain a certain closure. The modern nation-state is the architect of a variety of such exclusions comprising borders, suffrage, military service and far reaching welfare rights. However, the state's ability to maintain order and manage these tasks in turn pivots on the institution of citizenship. Only as a member can you *not* be excluded from entering your state's territory, and only as a member can you claim certain entitlements to goods and services from your fellow citizens that are crucial for shaping your life.

THE POLITICS OF PLACE AND ORIGIN

The above discussion concerning the language games of consent and representation suggests that the semantic field of our various concepts and their interconnection is powerfully influenced by the drawing of lines between inside and outside and between members and non-members. Furthermore, we found that the spatial and the ascriptive distinctions are interdependent, and thus powerfully shape our solidarities and define the distinct nature of the political game. But despite all of our efforts we could not provide a contradiction-free theoretical account which assigns a definite place to each concept and systematically includes the important connections between them. Finally, aside from cognitive issues, emotional attachments seem to play a decisive role, and no analysis can hope to be complete unless it pays proper attention to those factors. But what type of analysis could comprise all of these themes? Obviously, there is a need for the inclusion of some social-psychological approaches to unearth the logic of emotions.[30] Here the history of political thought provides us with some clues.

All of the above themes are already paradigmatically woven together in the myth of Er in Plato's *Republic*. Socrates speaks of the necessity to tell the members of the *polis*, a 'Phoenician tale', i.e. a story that is a 'myth'. Literally not 'true', it conveys a 'truth' which is not based on purely cognitive factors but which is designed to shape attitudes, and evoke feelings of pride and approval. It is this feeling of identity that allows the individual members of a body politic to act as a community, and to transcend the divisions and differences that are so obvious in everyday life. The *gennaion pseudos* – variously translated as 'noble lie' or 'pure fiction' – that Socrates wants to tell the rulers and ruled alike, is as follows:

> They are to be told that their youth was a dream and that the education and training which they received from us, an appearance

only; in reality during all that time they were being formed and fed in the womb of the earth, where they themselves and their arms and appurtenances were manufactured; when they were completed, the earth their mother sent them up; and so, their country being their mother and also their nurse, they are bound to advise for their good, and to defend her against attacks, and her citizens they are to regard as children of the earth and their own brothers.[31]

What is of interest for us is not Plato's teaching in general, or the specific role this 'tale' plays within the Republic in particular. For our purposes, it is the symbolic strategy that Plato uses, the form of symbolic representation which he hopes will ground a political community and which induces solidarity and motivates political action. For this purpose the myth of common descent is inculcated. But it is also a descent which is not lost in indeterminate time but rather one in which the land, the earth, serves as the common origin, as the mother of them all. As being a child is not a matter of choice, so being born into a community with all its attendant privileges and obligations is not like being part of a club or network.

It has always been considered odd that Plato, who riled at the poets and their false tales, chose at crucial points in his dialogues to utilise 'myths' in order to convey some of his most fundamental messages. Part of the solution to this puzzle might lie in the fact that these existential questions do not concern simple cognitive issues. If they are reduced to such, they quickly end in contradictions. Why, according to the logic of the liberal account, should the contract be insoluble? Why should future generations be bound by some voluntary undertakings that occurred several generations ago? If we were dealing with a genuine contract, such circumstances would in all likelihood allow for the invocation of the *clausula rebus sic stantibus* (changed circumstances which void obligations). Obviously, something more is at issue.

Such incoherence is then often quickly masked by an additional postulate, i.e. that of the inevitable rise of modernity. To that extent it is suggested that the non-cognitive problems to which some classical texts point, such as the Platonic tale above, are either simply denied or treated as overtaken by historical developments. As modern actors, we are supposed to have interests rather than attachments, we are to conceive of politics as an instrumental activity – properly defended as the paragon of rationality itself – and to be committed, if to anything at all, to progress and democracy as the end of history. Was not Plato, after all, a proto-fascist? And is not the call for a return to the classics part of a well-known conservative strategy? Have not the passions engendered by nationalism shown the dangers of admitting emotions into the public space? And, therefore, is not the only hope to push for universal rights and a new conception of cosmopolitanism? Has not the communications revolution, and the processes summarised by the concept of globalisation shown the atavistic nature of the state and traditional concepts of community?

Without examining the *prima facie* plausibility of these, admittedly, rather diverse claims, I only want to point out that they effectively prevent us from understanding ourselves and the world around us. This is not an advocacy for a return to a conservative tradition or canon (although 'progressives' have obviously similarly holy texts, and engage in precisely the same argumentative gambits), or to subject a political philosopher like Plato to the standards of a sophomoric debating team. It is rather a call for rethinking precisely those conventional wisdoms that served as the background for the above caricature. As a caricature it distorts, but it does so not by falsifying what is there, but by highlighting salient characteristics.

A quick look suggests: that the 'end of history'[32] has not arrived; that one of the most distinguished proponents of the rational choice approach warns us not to become 'rational fools' by making it appear that political choices are all like consumer choices[33]; that a strong advocacy of universal human rights still presupposes the framework of functioning states and societies; that fanaticism and irrationality are (unfortunately) not the exclusive preserve of people who have particularistic attachments (as the terrors waged for universalist goals of the liberation of mankind and the communist deliverance show); that emotions also form the basis of our moral judgements (to paraphrase Hume's argument about the importance of the sentiments of approval and disapproval);[34] that globalisation and modernity have not overcome the old identifications but rather stimulated the search for 'roots', for which (like it or not) nationalism still provides the most powerful idiom.

Should we not ponder more carefully Benedict Anderson's astute observation that there is in nearly every country the tomb of the unknown soldier, but none of the unknown Marxist?[35] Apparently, there is an appeal in nationalism that other ideologies have great difficulty in matching, even if it has nothing to do with primordialism as is sometimes suggested. After all, nationalism might be a thoroughly modern concept, although the invocation of a glorious past, the myths of common descent, and the particular importance of place utilise themes that have an ancient pedigree indeed, as the above Plato quote demonstrated. Nationalism though, is a phenomenon of modernity and not only in the sense of Gellner,[36] who emphasised the need for homogenisation of ancient ethnic and status differences in the wake of the industrial revolution.

Nationalism not only allowed people to leave their old station in life, to participate as enfranchised citizens in public affairs, but also made people more alike by subjecting them to a uniform educational discipline. Through national curricula and compulsory education, through the inculcation of national history, people not only acquired a new identity, far removed from that imposed by the traditional order, but they also became interchangeable exemplars of the standardised technical skills and know-how needed for mass production. Nationalism was for Gellner clearly a 'stage' in the development of societies, a development driven by the successive modes of production. Given the present circumstances of increasing interdependencies, it was a rather atavistic form of social and political organisation.

But such characterisations are problematic not only for the reason that attempting to map entire societies by the mode of production or stage of development is clearly obsolete as Offe and Preuss correctly pointed out.[37] Even more important for the purposes of legitimisation is that after the death of God – the traditional guarantor of order – 'the people' remain the only source of legitimacy. Thus 'the people' are not only the authors of their own lives and social arrangements, they are also, in a strange twist of the old sophist adage, 'the measure of all things'. The death of God made the problems of a theodicy, traditionally part of our religious understanding, part of the political discourse. It is certainly no accident in this context that Hobbes talked about the Leviathan as 'the mortal God'. Despite many deep and decisive differences between religion and our secular understandings formed by the Enlightenment, both presuppose that the

> destiny of mankind requires justification through the will of a

creator that binds humankind in a good order, whether divine or secular. This reference to the concept of political theology (or as it were, to the idea of immanent transcendence) may help us to understand the tension between the claim of the political order to be good and just, and the omnipotence of its sovereign – a tension which can only arise where we cannot resort to the authority of any external norms and principles of justice.[38]

We have traced some of these tensions in the above discussion concerning representation. But what needs further elaboration is the connection between this new understanding and the politics of identity to which it gives rise. Uncovering this connection will also help to explain why the nationalist call for a return to the roots resonates with, and is propagated most decisively by, intellectuals rather than by people who live the embedded life of their traditional communities. It is, therefore, a radical misunderstanding of this phenomenon to interpret it as an atavistic remnant of a time long passed. Rather, more like the modern revitalisation of religious fundamentalism, it is one attempt to come to terms with the twin problems of modernity: identity and authenticity.[39]

Only against the background of the modern preoccupation with the autonomy of the individual does the search for an authentic expression of 'the self' make sense. Only against the particular valuation of diversity, and the implicit understanding articulated by Herder that each distinct people had a historical mission, did the principle of 'self-determination' develop its explosive potential. Thus nationalism answers to a variety of modern problems in the construction of meaning.

Autonomy, the key to dignity in the modern world, requires authenticity; freedom depends on identity, and destiny on shared memory. So the desire to participate in a modern world of wide opportunities and technological expertise, requires the forging of separate moral communities with incommensurable and authentic identities. But if the secret of identity is memory, the ethnic past must be salvaged and reappropriated, so as to renew the present and build a common future in a world of competing national communities.[40]

By joining pre-modern ties and sentiments, characteristic of traditional *ethnies*, with the modern idea of popular sovereignty, nationalism provides an answer to the crisis of meaning engendered by modernity. Instead of the traditional myth of ethnic election by the deity, modern nations insist on the uniqueness and the value of their historical heritage. The concepts of identity, autonomy, authenticity, unity, and fraternity form part of a discourse that not only explains by means of conceptual elaboration and the narratives of myths and histories, but by expressive symbols and ceremonials that evoke the sentiments of self-worth and dignity.

Symbols and ceremonies have always possessed the emotive qualities described by Durkheim, and nowhere is this more apparent than in the case of nationalist symbols and ceremonies. Indeed, much of what Durkheim attributes to the totemic rites and symbols of the Arunta and other Australian tribes applies with far greater force to nationalist rites and ceremonies, for nationalism dispenses with any mediating referent, be it totem or deity; its deity is the nation itself. By means of the ceremonies, customs and symbols every member of the community participates in the life, emotions and virtues of that community and through them, re-dedicates him or herself to its destiny. By articulating and making tangible... the concepts [sic!] of the nation, ceremonial... symbolism helps to assure the continuity of an abstract community of history and destiny.[41]

While this passage points to the importance of the symbolic dimension of politics (adumbrated in the discussion of symbolic representation above), we should also realise that the emotions of dignity and self worth can of course easily derail and lead to the denigration of others, even to hatred, for which we all, unfortunately, can readily cite examples. Nevertheless, without feelings of self-worth and dignity it is hard to imagine that our most basic moral assessments could function. The connection between the lack of feelings and the baseness of character, between the irresponsibility of a person, and his or her lack of self-esteem, are well established by psychology. Even if one has to be particularly careful with inferences from individual to collective behaviour, there is no doubt that the moral discourse treats collectivities analogous to individuals, and that feelings of self-worth and appreciation are constitutive of the definition of individual and collective roles. Thus, when Robert Kennedy, during the Cuban missile crisis, rejects the advice of a surprise air strike on the basis that 'we' are not like Tojo, he not only shows the relevance of these considerations even for 'high politics' and for crisis situations, but he also demonstrates that decision-makers do make inferences jumping from the individual to the collective level. It is here that political theory most clearly needs help from psychology and social-psychology, not only for cognitive issues, but also to clarify the role of emotions in shaping thought and perception.[42]

CONCLUSION

The perspective outlined above may leave us with more questions than answers. It began with an inquiry into the question of legitimacy via the traditional liberal arguments for consent as the basic foundation. It soon had to expand its scope by examining the language game of representation. By inquiring into the meaning of the term 'the people', we finally had to give up hope of finding a fixed referent, but were directed to the activities of drawing conceptual boundaries. It was in this context that the importance of place and origin emerged for filling in the cipher of 'the people'. But seeing that our theoretical terms are part and parcel of systems of signification, which we construct and deconstruct by tracing the historical developments of various conceptual linkages and mutual referrals, we must not believe that everything is possible.

The modern nation might be an imagined community and, thus, a product more of our abilities of symbolic manipulation than of natural facts, but this imagining takes place in an environment which is not empty. In it we find the sediments of former imaginative acts that sometimes occupy central places. Besides, 'imagining' a community is not like wishing it into existence. It entails a collective enterprise that has to give rise to particular disciplinary understandings as well as to settled practices. The notion that this activity of imagining is somehow preordained by the course of history is one of the phantasms of modernity. The critical remarks above were not only designed to dispel such private flights of fancy, but also to show that nationalism is a specific response to the strains of modernity. To that extent it is neither atavistic, nor is it likely to be simply overcome. Two of the other major contenders, religion and human rights, can answer some of the existential questions adumbrated above, but neither of them explicitly addresses the social configurations that have attained such prominence in modernity. Even the Islamic revival takes place *within* the organisation of the state system, and if anything can be learned from the fate of universal religions, it is that they are not able to organise 'mankind' as an undifferentiated totality.

Similarly, the secular attempt of claiming universal status for human rights does not do much better, partly because this discourse is constituted by a rather disparate amalgam of various traditional strands of thought. There is above all an uneasy link between the natural law tradition, which relies on some intuition into what *is* right, and the notion of subjective *rights*, i.e. claims that have to be respected even if the right-bearer does the 'wrong' thing. To that extent, the notion of 'right' is a convenient cipher to cover up significantly different meanings. But it is not surprising that conflicts appear when e.g. a group wants to punish a writer who has done the wrong thing in exercising his rights, while others maintain that such a reading not only smacks of fundamentalism but negates the very notion of having a right. Even if we all agreed that the second version is the (politically?) correct one, our troubles are not over. Any subjective right is a socially respected claim which, in turn, presupposes a functioning society. To that extent it seems not only practically difficult, but logically impossible to found a society on the assertion and the exchange of rights. We either slide back to some intuitionist version of the Laws of Nature and some concomitant notions of what *is* right, or we have to face the problem of contract theorists, that without 'the people' and the establishment of shared conventions of what constitutes a contract, no such undertaking can get off the ground.[43]

The last point speaks to the difficulties in consent theories as elaborated above and to the contemporary confusion of making justice conceived as fairness and as 'taking rights seriously' the paramount political problem. Aside from the fact that other important problems of politics are thereby eliminated, most theorists in this vein admit, when pressed, that their speculation is relevant only when we already possess well-established states, or at least civil societies. Furthermore, there is a whole literature on the limits of law that casts serious doubt on the proposition that a society can survive and function when it is based solely on the assertion of subjective rights.

In short, the value of human rights consists neither in the conception that we find some inchoate international authority which enforces them, or that the universalist program transcends the existing social arrangements of nations and states, but, rather, in the fact that human rights have become part of the legitimising discourse *within* each state. It is to those rights that people can appeal in appraising and (de)legitimising political acts. To what extent this discourse will be successful in replacing other notions of right and wrong, which have their roots in other intellectual traditions, remains to be seen. Given that increasingly particular political agendas are pursued by western states under the flag of human rights, it is not surprising that such policies engender resistance from leaders and various strata in non-western countries, their popularity among certain local groups notwithstanding.

Thus the dream of a world culture, and the victory of universalism against traditional parochialism seems largely mistaken. If memory is central to identity then the futuristic fantasies of the technotronic age are unlikely to be able to address this problem of identity formation. The fact remains that the elements that would help us with building new identities are actually largely appropriated by specific cultures and traditions. Nowhere does this become more obvious than in the case of Europe, where even a common cultural awareness is unable to create a viable new political identity. If this is so, then chances for a global revolution of consciousness are dim indeed. As Anthony Smith points out:

> The packaged imagery of the visionary global culture is either trivial or shallow, a matter of mass commodity advertisement, or it is rooted in existing historical cultures, drawing from them whatever meanings and power it may derive... For a time we may be able to get by and invent traditions and manufacture myths. But if myths and traditions are to be sustained, they must resonate among large numbers of people over several generations, and this means they must belong to the collective experience and memory of particular social groups. So new traditions, too, must be culture specific: they must be able to appeal to and mobilize members of particular groups while excluding, by implication, outsiders, if they

are to maintain themselves beyond the generation of their founders.[44]

Thus, instead of the ideology of modernity which interprets history as a development to ever more inclusive forms, thereby realising the cosmopolitan ideals of mankind, the truth of the matter is that patterns of integration as well as assertions of particularity are part of our predicament. No purpose is served when, instead of investigating these two equally important phenomena, one is treated as normal, while the other is relegated to oblivion. As its name suggests, (and this chapter has argued) *politics* seems thoroughly partial.

NOTES

1. See Yosef Lapid and Friedrich Kratochwil, 'Revisiting the National: Toward an Identity Agenda in Neorealism?' in Yosef Lapid and Friedrich Kratochwil (eds.), *The Return of Culture and Identity in IR Theory* (Boulder, CO.: Lynne Rienner, 1996), ch. 6.
2. David Held, 'Democracy, the Nation State, and the Global System' in David Held (ed.), *Political Theory Today* (Stanford, CA.: Stanford University Press, 1991), p. 222.
3. Lea Brilmayer, 'Consent, Contract, and Territory', *Minnesota Law Review* (Vol. 74, 1989), p. 6.
4. For an interesting case in international law which had to deal with these issues see *Case Concerning the Barcelona Traction Light and Power Co.* (Begium v. Spain), 1970 I.C.J. 3
5. John Locke, *A Second Treatise of Government*, § 121, in Peter Laslett (ed.), *Two Treatises of Government* (New York, NY: Mentor, 1963), p. 393f.
6. *Ibid.*
7. *Ibid.*, §122, p. 394.
8. *Ibid.*, e.g. §98 and § 99
9. *Ibid.*, §98, p. 376.
10. *Ibid.*, § 140, p. 408.
11. Don Herzog, *Happy Slaves: A Critique of Consent Theory* (Chicago, IL: University of Chicago Press, 1989), p. 185.
12. *Ibid.*, p. 194.
13. On this point, see my 'The Limits of Contract', *European Journal of International Law* (Vol. 5, No. 4, 1994), pp. 465–91.
14. Locke, *Second Treatise*, § 128, *op. cit.*, p. 397.
15. See e.g. his article 'Justice as Fairness: Political not Metaphysical' in *Philosophy and Public Affairs* (Vol. 14, Fall 1985), pp. 223–51.
16. David Hume, *The History of England*, New Ed., 6 Vols. (Boston, MA.: Little, Brown, 1872).
17. Max Weber, *Economy and Society* (Berkeley, CA.: University of California Press, 1978).
18. Hanna Fennichel Pitkin, *The Concept of Representation* (Berkeley, CA.: University of California Press, 1972), p. 87.
19. James Madison, Federalist 10 in *The Federalist Paper*, edition by Clinton Rossiter (New York, NY: Mentor, 1961), p. 82.
20. *Ibid.*, p. 84.
21. For a fundamental discussion of the role of emotions for politics, and the 'condensation symbols' that active these emotions, see Murray Edelman, *The Symbolic Uses of Politics* (Chicago, IL: University of Illinois Press, 1967), *passim*.

22. Philip A. Griffith and Richard Wollheim, 'How can one Person Represent Another?', *Aristotelian Society* (Suppl. Vol. XXXIV, 1960), p. 189.
23. See Ronald Dworkin, *Taking Rights Seriously* (Cambridge, MA.: Harvard University Press, 1978); and Ronald Dworkin, *Law's Empire* (London: Fontana Press, 1986).
24. See, for example, William Harris, *The Interpretable Constitution* (Baltimore, MD.:Johns Hopkins University Press, 1993).
25. As quoted in Pitkin, *op. cit.*, p. 30.
26. Thomas Hobbes, *Leviathan*, *op. cit.*, ch. XVIII, p. 229.
27. On this point see also William Zartman, 'Self and Space: Negotiating a Future from the Past' (Mimeo 1996).
28. Rogers Brubaker, *Citizenship and Nationhood in France and Germany* (Cambridge, MA.: Harvard University Press, 1992).
29. On the argument about the enduring character of ethnic attachments, see the collection of essays by Walker Connor, *Ethnonationalism: The Quest for Understanding* (Princeton, NJ: Princeton University Press, 1994). The term 'primordial feelings' was first used by Edward Shils, however in a context which suggested that there exist a variety of such 'primordial' attachments; see Edward Shils, 'Primordial, Personal, Sacred, and Civil Ties', *British Journal of Sociology* (Vol. 17, 1957), pp. 113–45.
30. For a brief introductory discussion of the issues involved see Robert Solomon, *The Passions : Emotions and the Meaning of Life* (Indianapolis, IN.: Hackett, 1994); see also Amelie Oksenberg-Rorty (ed.), *Explaining Emotions* (Berkeley, CA.: Campus, 1980).
31. Plato, *Republic*, Bk. III, Trans. B. Jowett (New York: Achor Books, 1973), p. 105.
32. For the 'end of history' debate see Francis Fukuyama, *The End of History and The Last Man* (New York, NY: Free Press, 1992).
33. See Amartya Sen, 'Rational Fools: A Critique of the Behavioral Foundations of Economic Theory', in Jane Mansbridge (ed.), *Beyond Self Interest* (Chicago, IL: University of Chicago Press, 1990), ch. 2.
34. See David Hume, *Treatise on Human Nature*, Bk. III 'Of Morals', in *Hume's Moral and Political Philosophy*, edition by Henry Aiken (Arien, CT: Hafner, 1970).
35. Benedict Anderson, *Imagined Communities* (London: Verso, 1983), p. 10.
36. Ernest Gellner, *Nations and Nationalism* (Ithaca, NY: Cornell University Press, 1983).
37. Claus Offe and Ulrich K. Preuss, 'Democratic Institutions and Moral Resources' in Held, *op. cit.*, p. 144.
38. *Ibid.*, p. 147.
39. Elie Khedourie has carefully examined the sources of nationalism and their connection with the specific problems brought about by the crises of modernity. Elie Kedourie, *Nationalism*, 4th ed. (Oxford: Blackwell, 1993).

40. Anthony D. Smith, *Nations and Nationalism in a Global Era* (Cambridge: Polity Press, l995).
41. Anthony D. Smith, *National Identity* (Reno, ND.: University of Nevada Press, l991), p. 78.
42. For an important discussion of 'subject referring' emotions and their role in identity construction see Charles Taylor, 'Agency and the Self', in Charles Taylor, *Human Agency and Language* (Cambridge: Cambridge University Press, l985), Pt. I.
43. For a further elaboration of this point, see my *Rules, Norms, and Decisions*, (Cambridge: Cambridge University Press, 1989), ch. 6.
44. Smith, *Nations and Nationalism, op. cit.*, p. 23f.

9. International Relations Theory and The Fate of The Political
R.B.J. Walker

This chapter pursues an argument that is at once simple and grandiose. It suggests that far from being a backwater that can be ignored in favour of currently fashionable forms of literary, cultural, sociological, ethical or economic theory, or that can be left to reproduce the tired clichés of Cold War social science for the declining cadre that remains loyal to them, the theory of international relations confronts an especially crucial and difficult range of problems for our established accounts of *political* life. Of all the established disciplines of the modern social or human sciences, it seems to me, only anthropology comes anywhere near it for the challenges it poses to our inherited accounts of what it means to understand or intervene in human affairs. Recent critical engagements with the theory of international relations are important not just because they have drawn attention to the (fairly obvious) poverties of positivism, the neglect of economic determinations, the complexities of states, the discursive constructions of nationalisms and enemies, the marginalisations of gender and culture, and so on, though these and other engagements have had a wide range of constructive effects. They are important primarily because they force us to come to terms with questions about the character and location of politics given the uncertain status of the sovereign state as the primary constitutive principle governing what and where modern politics is supposed to be.[1]

Given the limited length of the chapter, I engage here with the simple rather than the more grandiose aspects of this argument.[2] The engagement is also rather indirect, focusing on a recent brief, but usefully sharp commentary by Kal Holsti on the status and possible future of international relations as an academic discipline. Over the past decade or so, a considerable number of vibrant literatures have emerged in reaction to what many have seen as the limited intellectual resources of this discipline.[3] Indeed, theories of international relations after the Cold War seem to have sprouted with all the vigour, variety, and, some appear to fear, toxicity of mushrooms after rain. Contrary to Holsti, who expresses sceptical sentiments that are widely shared among more established theorists of international relations, I want to suggest that recent critical openings should be interpreted positively. But I also want to suggest that many of the critical literatures that worry commentators like Holsti are themselves in danger of marginalising questions about politics quite as much as the traditions they have sought to challenge. Whatever short-term stimulations they might inspire, recent attempts to develop more innovative – more critical, more gender/culture/history-sensitive, more ethically and developmentally appropriate, less structurally determinate – forms of international relations theory will have to pay more attention to the political implications of their innovations. And by this, I mean quite a bit more than simply paying attention to the relationship between theory and practice or theory and policy as this has been understood by those impatient with recent preoccupations with theoretical controversy of any kind.[4]

To put this rather differently, I want to suggest that it is about time that theorists of international relations take their traditional patron saint – Thomas Hobbes – rather more seriously than they have. Instead of attending to the rest of what I say here, in fact, one might usefully spend some time thinking about why Hobbes, like so many contemporary theorists of international relations, was so obsessed with the politics of language, and why the established conventions of international relations theory have managed so successfully to ignore what he had to say, not only about language, but also about the status of sovereignty as the necessary condition of the possibility of a modern politics. Contrary to many voices proclaiming that we have now paid more than enough attention to metatheoretical matters, in effect insisting that we should simply accept the framing of all metatheoretical options bequeathed to us by Hobbes and his successors in the received canons of modern thought, I want to counsel quite the opposite. For if there is some plausibility to the emerging consensus that the old statist view of political life associated with Hobbes and his successors offers only a limited or even completely misleading view of contemporary trends, then the assumptions that permitted the production of this conventional view are unlikely to be of much help in thinking about those trends. In many ways, in fact, Hobbes's assumptions may be more deeply rooted in the contemporary political imagination than either the state or the states system those assumptions are presumed to endorse.

Put very briefly, the general argument that informs this chapter rests upon two sets of claims. First, that much if not most of what is at stake in a wide range of contemporary debates about (at least) (i) the status of international relations theory; (ii) its relations to prevailing traditions of political theory, political economy and ethics; and (iii) the status of various claims about emerging forms of 'globalisation' and 'world politics', can be read as variations on a question about the location and character of what we mean by the political under contemporary conditions. Second, that contemporary transformations in the location and character of the political can be read as revalorisations of relations of space, time and identity that the modern world has come to treat as resolvable through principles of a sovereign and self-identical subjectivity in general, and the sovereign state (and by derivation, the statist-nation) in particular.

Modern concepts of sovereign subjectivity, I want to insist, are quite misleadingly read as extremes – as a matter of fragmentation and anarchy, as the more conventionally realist or utilitarian theorists of international relations would have it, or as a regulative but elusive aspiration for autonomous individuality, as the (liberal) political theorists and nationalists might prefer. They are, rather, claims about the possibility of a middle ground between extremes. However else they might be read in relation to, say, the functional requisites of global capitalism, the structures of geopolitics or colonialism, the constitution of genders or the legitimation practices of modern states, claims about modern sovereign subjectivities express a settlement with cultural and metaphysical options, a settlement that has in turn crucially shaped both the possibility and the sustainability of politics in the modern state.[5] It is the plausibility of this settlement, this reconciliation of the apparently irreconcilable, that is in question. For a middle ground between implausible extremes is, and has long been, a tenuous place to be. As Hobbes recognised perhaps more clearly than most, it certainly does not offer the firm foundations that so many have come to confuse with modern reason.[6] Moreover, the most immediately attractive route to safer ground – to the extremes of unity or diversity that are supposedly mediated by the great unity-in-diversity of modern sovereign subjectivities – inevitably turns out to be a chimera. To call sovereignty into question, as perhaps even a majority of international relations theorists are now prone to do to some degree, is to provoke questions about much more than the status of the modern state.

The first set of claims leads to a range of concerns that are usually identified as the proper responsibility of political theorists. The second set seems to require that I be branded, according to the current conventions, as some kind of postmodernist. As I understand them, however, the implications of these two sets of claims are that (i) anyone who knows for sure what the proper responsibility of the political theorist should now be is in very serious trouble, and (ii) that those who take delight in dividing defenders of modernity from celebrants of postmodernity reveal only their own fondness for specifically modern techniques of classification and order. In fact, one of the dangers I want to address here, at least in part, is that having begun to open out in so many fertile directions, the theory of international relations is still susceptible to the seductions of a modern conception of political subjectivity of precisely the kind that enabled and sustained the most reified forms of international relations theory that many of us hoped it might be possible to jettison.

At some point, this general argument will have to engage with a wide range of empirical trends. For the moment, however, the greatest difficulties seem to me to relate to the categories and discourses through which one might propose to decipher empirical trends or make judgements about their relative importance. Too much energy is still wasted on pre-Kantian squabbles over whether empirical research is somehow desirable or undesirable, and not enough on examining the conditions under which *any* claims to knowledge get to be made or sustained.

In one form or another, the general argument informing this chapter is being pursued, also rather indirectly and speculatively, by a wide range of contemporary literatures across many fields of scholarship, policy analysis, journalism, and even popular culture. Claims about globalisation have become especially important in this context over the past few years.[7] Debates about international relations theory are, however, significantly privileged in this context both because they, perhaps more than any other arena of contemporary thought, are always supposed to be aware both of the *limits* of a modern politics expressed by the principle of state sovereignty and of the consequent contradiction inherent in *any* claim to a politics of 'the world', that is, a politics that exceeds the legitimate authority of sovereign territorial jurisdictions. Indeed, it sometimes seems that the more empirically interesting the claims made about globalisation, world politics, or whatever emerging patterns of economic, technological and sociocultural determination on a world-wide scale are to be called, the less interesting they are as analyses of changing patterns of political community and legitimate authority. Supposedly political economies, sociologies of world systems, and claims about globalisation still seem to be especially susceptible to the silent presence of statist accounts of political life even as they rightly analyse the embeddedness of states in broader and more complex global/local processes.

International relations theory, then, is clearly changing very rapidly under a wide range of pressures. For some, this is a matter for considerable concern, while for others it is cause for celebration. Both judgements have merit. In this chapter, I want to move from one of these judgements to the other: first, to consider some reservations about contemporary trends posed by those like Holsti who look back on earlier forms of international relations theory with a degree of (qualified) nostalgia, and second, to suggest that there are other ways of understanding what is at stake in these reservations, and especially in the claim that the theory of international relations as we have known it is now in some kind of crisis. In largely affirming a more celebratory account of recent developments, however, I want to underline the seriousness and difficulty of the questions that arise once recent critical literatures in the specific field of international relations are also read more broadly as critiques of modern conceptions of the political.

DISCIPLINE IN CRISIS/THEORY OF CRISIS

Holsti has articulated an especially cogent version of the claim that various theoretical trends have led to a serious and undesirable crisis in the theory of international relations.[8] He discusses four possible explanations for, or at least symptoms of, such a crisis: an undermining of theoretical authorities by 'critical' and 'postmodern' theorists; a lack of isomorphism between prevailing theoretical traditions and the emerging contours of the post-Cold War era; a sense in many quarters that there is no longer a commanding core problem to be addressed, so that questions about epistemology, ontology and the relation between theory and practice have come to replace a more appropriate concern with substantive problems; and a suspicion in other quarters that the core problems of international relations, and especially the incidence of interstate war, have been solved. I share Holsti's apparent scepticism about the last of these but have a rather different sense than he does of what is at stake in the other three.

Holsti also canvasses three possible scenarios for the future development of the discipline. First, 'oblivion': the disintegration of the discipline through a 'scholasticism' concerned about 'the innumerably contested ways of *how to think* about a subject matter' rather than about 'the subject itself'[9]; or through a concern with proliferating forms of identity politics; or through a forgetting of important traditions in the name of opening up 'thinking space;' or through a rejection of the constitutive distinction between domestic politics and international relations. Second, an uncivil and unscholarly war between self-righteous theoretical factions leading nowhere. And third, the emergence of some new consensus (perhaps an international political economy; or a concern with problems of global governance, system transformation or emancipation of some kind; or something new emerging from societies that have not dominated the field so far). Several other commentators have recently made parallel observations, usually as a prologue to a series of recommendations for an alternative form of analysis, but Holsti effectively captures a mood that is widely shared, and poses his critique in a manner that is both blunt and constructively ambivalent.

Holsti's conclusion is cautiously optimistic:

> [w]hile international theory will never be a unified field because its normative sources differ... we might envisage a field that matures in philosophical depth and self-consciousness, but which is also aware of the contributions that a variety of intellectual perspectives, research strategies, and methodologies can make to greater understanding and which, above all, focuses on actual political dilemmas in the world.[10]

For all that this conclusion moves in an open and pluralistic direction, however, it is also clear that Holsti is deeply concerned about the loss of a traditional consensus and about the disruptions wrought by recent critics and the proliferation of research interests. It is this tension in the analysis that I find interesting. It is not difficult to agree with his diagnosis at many points, and not least with his concern with the consequences of theoretical self-righteousness, which is certainly visible among established and emerging scholarly traditions alike. But it is precisely where he sees some of the greatest dangers that I tend to see some of the greatest opportunities now open to us. This (perhaps only slightly) more positive reading of recent controversies and trends requires a rather different perspective on what one thinks international relations is, what a crisis is, and how claims about both theories of international relations and about crises are precisely expressions of practices that Holsti dismisses as scholastic rather than substantive. Holsti is rather more impressed than I am by claims about the 'perennial' quality of established traditions.[11] I am rather more impressed than he by the intimate, historical and mutually constitutive relationship between what appear to be substantive realities and problems and the procedures through which we claim to know about and act upon those realities and problems. These differences, I think, open out into quite different readings of what is at stake in debates over the future of this discipline.

In any case, Holsti's analysis can be read as suggesting that the question of whether there is indeed a crisis in international relations theory is less interesting than the question of what is at stake in the claim that there is such a crisis. What, it leads me to ask, are the conditions under which it has become possible to make such claims? How does the claim get to be articulated, for what purposes, by whom and for whom? And why, beyond the immediate self-interests of those with a stake in a particular profession, should anybody care?

The primary condition under which recent controversies about the state of the discipline have become possible, I believe, involves the way in which the theory of international relations already tells us what a crisis looks like and how one is supposed to respond to it. It does so in three interrelated senses. First, the theory of international relations is already constituted as a theory *of* crisis, though not, crucially, simply a crisis in what is called international relations. Second, as a theory of crisis, it is widely perceived to be itself *in* crisis, to have reached the limits of what it is able to say about the character, location and management of crisis in modern political life. Third, from some quarters at least, the very notion of what one means by a crisis is also said to be in some trouble, for reasons that are directly linked to the inadequacies of the theory of international relations as a theory about the limits of modern politics.

International relations theory is a theory about crisis, about the limits of the normal; this part of my argument simply affirms what I take to be many of the most conventional assumptions about what international relations theory is and does, though it does so, I believe, without buying into any notion of a reified perennial tradition. In one crucial respect, however, it turns these conventional assumptions on their head: in this reading, international relations theory expresses an account of how the world *must* be, given the claims of the modern state; for all the rhetorical resort to claims about political realism, common sense and policy relevance, the theory of international relations is primarily a normative enterprise.

Many now claim that this theory has reached some sort of limit in its ability to explain and express the limits of the normal. Again, this is to affirm a claim that can be heard from a great many quarters; it speaks, for example, to Holsti's observations about a widely shared concern about a lack of isomorphism between prevailing traditions and contemporary circumstances.

And finally, many of the great conceptual and practical difficulties of our age, and not least our difficulty in expressing and explaining phenomena that once fitted quite comfortably into the received categories of international relations theory, are linked to a profound sense of the inadequacy of modern accounts of time, change, history, evolution, revolution, development and so on, as ways of thinking about present trends and future possibilities. This is where the conceptual terrain becomes more difficult, but also, I think, more interesting. It is here that some apparently arcane, and to some observers quite irresponsible, trends in contemporary social and political theory impinge most threateningly on established conventions and categories of analysis. It is here that Holsti and I most clearly part company.

Given the extent to which specifically modern accounts of what a crisis is have close historical and logical connections with the rise of the modern state and states system in early-modern Europe,[12] the intimate interplay between these three senses in which contemporary claims about crisis are played out should not be surprising. But given the trivialisation of so many of the key issues that are at stake, I do think it is worth reminding ourselves of legacies we have inherited from the early-modern period, legacies that can be ignored now only at considerable intellectual and political cost. To put it rather bluntly, considerable nonsense has been uttered in many recent debates about the future of international relations precisely because they continue to be framed on the assumption that specifically modern claims about a sovereign subjectivity can be sustained as the final ground on which debates about knowledge and power can be resolved. For better or for worse, this assumption seems to me to be implausible. It is at least as implausible, I believe, as the cosmological, cultural and political codes of 'the Great Chain of Being' were at the time when the modern state, the modern states system, and the modern conception of individual subjectivity emerged out of the ruins of empire, feudalism and the Thomist synthesis. Whatever else it might be, the theory of international relations expresses the limits, the point of crisis, in this specifically modern account of what it means to be a subject, a citizen, a human being. To suggest that this particular way of handling the limits of modern political possibilities is in crisis is presumably to say something that is far from trivial, and of interest far beyond the professional discipline of international relations.

This is not to say that the attractions of modern claims about autonomy, self-government, national self-determination, and so on, are simply going to disappear, any more than the hierarchies of the Great Chain of Being simply disappeared once Descartes, Hobbes and the rest had constructed a new way to speak of being human by standing on the shoulders of the nominalist theologians. Modern conceptions of autonomy and sovereign subjectivity are, I am prepared to admit, a lot more attractive than many other options. But modern conceptions of sovereign subjectivity work within modern accounts of the sovereign state, and the sovereign state, as we all know far too well, has always had a precarious relationship with other sovereign states. The niceties of political theory are quite inseparable from the nastiness of international relations, and vice versa. Machiavelli's dagger still splits the heart of the modern world and reminds us of the hypocrisies shrouded in our fondest hopes.

A THEORY OF CRISIS IN CRISIS?

In order to get some purchase on claims that international relations theory is in some kind of crisis it is necessary to work towards at least some clarity about what 'international relations theory' is taken to be and what it means to be in a condition of crisis. In this context, two

IR and the Fate of the Political

rather different understandings of international relations theory come into play, depending on whether one treats it as an attempt to explain phenomena or as an expression of the phenomena it claims to explain. The latter is, I believe both more accurate and more interesting. Both understandings, however, lead to suspicions that whatever is at stake in claims about a crisis in the theory of international relations goes much beyond the specific substantive concerns of this theory. While affirming these suspicions, however, I also want to insist that the theory of international relations expresses distinctive limits on what can be achieved by going beyond its conventional boundaries. The very frailty of established conventions can easily seduce critique away from the significance of a discipline that has such strength precisely because it can easily appear to be so frail.

In one understanding, the theory of international relations appears primarily as part of the modern social sciences; as an academic discipline shaped especially by the (primarily 'liberal') ontological assumptions and (primarily empiricist) epistemological practices of modern social science. It is this particular understanding that tends to predominate in the United States, where much of the historical and philosophical constitution of modern politics tends to receive especially scant attention. The notion of a crisis arises in this context mainly in relation to widespread scepticism about those assumptions and practices (as with the influence of ontological claims about histories and plural identities and epistemological claims about interpretation and poststructuralism). It also appears in relation to the apparent collapse of disciplinary categories that had congealed partly in turn-of-the-century Europe and partly in the Cold War United States. In this context, a crisis in international relations theory is identified in relation to much broader intellectual trends that have affected all forms of scholarly enquiry to a greater or lesser extent. Consequently, it can be made to seem as if there is little, if anything, about the specificities of international relations theory that are of pressing interest. Indeed, the way is wide open to colonisation by the utilitarian individualism that seems to have been so persuasive in other fields of social inquiry. Social science is social science, and its methods and metaphors, it is said, can be applied universally. Likewise, more general critiques of what social science has come to be can be drafted in from the philosophy of social science, philosophy of language, cultural studies and all the other social science disciplines.

The usual shorthand used to describe specifically international relations scholars who are persuaded of the seriousness of a general rift with these established conventions involves claims about 'postpositivism' and the multiple varieties of 'critical theory'. The primary ground for debate tends to be epistemological in character, and at a broad rhetorical level at least, it is not difficult to identify recurring laments about lapsed standards of objectivity and the looming threat of relativism as the lowest common denominator informing claims about how international relations theory is indeed in crisis and that a Restoration is urgently required to restore peace, good order and reliable knowledge. The heart of the issue here, it is usually claimed, is whether the bulwarks that have been erected to protect us from the effects of Nietzsche, the neo-Kantians and the assaults on the great Newtonian certainties about a century ago can still be shored up despite the flimsy materials available to hold the dykes against the encroaching subjectivists, poets, and similar undesirables.[13]

Much as one might want to believe in the great socio-scientific hope that Nietzsche and the rest simply did not happen, I find it very difficult to resist the view that a certain level of honesty is called for here. The accounts of a proper social science that congealed in the 1950s, and the standards of scholarship these accounts continue to legitimise, are very clearly in trouble. Something might well be salvageable from them, but if international relations theory is to be understood as part of the broader project of modern social science, it seems rather pointless to elaborate some notion of a crisis in it without coming to terms with broader patterns in, to invoke Weberian/Machiavellian imagery, the fate of western or modern Reason.

In any case, whatever one's response to debates about epistemology, the fragility of the categorical distinctions among the contemporary social sciences is apparent to almost everyone. It has been apparent to some degree for most of this century. The continuing resort to familiar conventions in this respect is to be explained more by the sociology of knowledge and scholarly institutions that by any sustained defence of their continuing intellectual legitimacy.

In a second understanding, international relations theory appears more as a part of the world it seeks to explain or understand. The key issue here involves the claim that the world is changing in crucial ways, usually ways in which words like 'interdependence' and 'globalisation' are found useful and the principle of state sovereignty receives a daily funeral oration. In its simplest form, established theories are said to be inadequate to the task of explaining the new reality, whatever that is. Much more interesting, I think, is the claim that international relations theory is an expression of an historically and culturally specific framing of the world. Stanley Hoffmann once underlined the obvious point that international relations theory has to be understood in large part as a specifically American enterprise; I would want to underline the even more obvious point that it is an enterprise of the specifically modern sovereign state. My own view, in fact, is that there is really very little in the modern theory of international relations that cannot be extrapolated fairly straightforwardly from the way in which the claim to state sovereignty works to resolve all contradictions of unity and diversity in space and time upon a particular territory and a specific subjectivity. But whether seen as an increasingly feeble explanation, or as I would prefer, a decreasingly persuasive legitimation practice of a world of sovereign states, claims about a crisis in international relations theory clearly refer to something much broader than the practices of a specific academic discipline.

Again, to make claims about a crisis in international relations theory is to refer to problems that go well beyond the ambitions and achievements of a specific discipline. If state sovereignty is somehow in trouble as the great constitutive principle of modern political thought and practice, then any crisis in international relations theory can only be one expression of a more profound crisis in modern conceptions of political life in general.

Both of these identifications of what international relations theory is lead to the conclusion that any debate about whether it is or is not in crisis must be situated in some broader context. They can lead one away to other broader, and arguably more important literatures, such as to the great cultural industries recently devoted to famous French thinkers, for example, or to trendy debates about identity politics, feminism, postcolonialism and so on. And to a very large extent, such a broadening of horizons is indeed necessary if only as a reminder of the extraordinary parochialism, and culture/gender specificity, of the traditions and literatures that have come to define what international relations theory is supposed to be about.

But this would be to risk underestimating the specific importance of international relations theory both as a key site of modern social science and as an expression of the limits of a politics grounded in claims about state sovereignty. For the theory of international relations is not just another factory in the great industry of social science. Nor is it a field of inquiry concerned with just another aspect of politics. While the two primary accounts of what the theory of international relations is can lead to demands that one broaden one's horizons to examine some wider crisis in western or modern Reason or the plausibility of the claim to state sovereignty, they also lead to a renewed focus on what work the theory of international relations, understood in both of the above senses, does in those broader contexts. As many recent commentators of a 'critical' persuasion have come to argue, this work is precisely concerned with the setting of limits, that is, with those situations in modern life, in modern reason and in modern politics, in which crisis is the norm and not the exception.

This is, I want to suggest, the most salient aspect of any characterisation of what the theory of international relations that is said to be in crisis is supposed to be. International relations theory is *already* constituted as a theory of crisis. It expresses a specific account of what it means to be in crisis. To suggest that a theory of crisis is in crisis is to claim something that is especially disconcerting.

So what then is meant by crisis in this context? Here grand philosophies of history are leavened with occasional platitudes about dangers and opportunities. Beneath competing claims about the banalities and bankruptcies of the discipline lie grand debates about the end or the tyranny of History, about the fulfilment or fragility of teleologies of Progress and Development. The conceits of the Cold War, or of the Enlightenment, are said to be especially dubious. Scholars of a 'critical' persuasion, for example, are often said to have 'rejected' such conceits, as if anyone can so simply reject cultural assumptions that are constitutive of what it has been possible to become as a scholar or a person.

The generalised sense of endings or bankrupt traditions, however, is only part of what is involved in referring to a crisis in international relations theory, not least because it puts into effect a familiar story about an epoch of normality that has now been ruptured, whether by revolution (some neo-Kuhnian paradigm shift) or some more gradual evolution. Once upon a time, this story goes, there were clear standards of objectivity, or good foundational principles of Enlightenment Reason. Now these have been dissolved or deconstructed. The narrative is often coded as the move from modernity to postmodernity, and it is here that laments about relativism have been most insistent. Indeed, it has become almost impossible to engage in any form of reasonable discussion about ontological or epistemological questions in large areas of contemporary scholarship, precisely because the supposed move from modernity to postmodernity has been so overcoded and willfully misconstrued as one from an era of foundational principles to one in which the very possibility of such principles has been thrown into question.

Scholars engaging with the theory of international relations should be especially aware of the crudely rhetorical status of such claims. At least here, one would have thought, it might have been possible to take the implications of claims about state sovereignty with some degree of seriousness. For the most basic assumption of international relations theory, that of the centrality of the sovereign state as the locus of legitimate authority, is precisely dependent on the knowledge that there is no rational or natural ground upon which to found legitimate authority. From Hobbes to Schmitt, sovereignty is understood to be constituted through an ultimately arbitrary act. From Descartes to Kant to contemporary empiricisms, claims to knowledge are affirmed against a constitutive scepticism. God was already mouldering in the grave by the early-modern era, and many of those who tried to put substitutes in his place were quite aware of the artifice necessary to persuade citizen-subjects that Reason and Law could work as well if not better than divine edict. The history of modern politics is a story of the success or otherwise of principles and institutions that have been constructed upon an arbitrary act of will, on the ultimate authority of the sovereign that can decide the exception to every general rule. The move from modernity to postmodernity cannot usefully be construed as one from foundationalism to relativism because the supposed foundations of modern life have been unhinged from any mooring for a long time.

This is why Hobbes, the great nominalist student of rhetorical strategies of persuasion, offers such a suggestive articulation of what is at stake in the theory of international relations, even though he says very little about international relations as such and explicitly denies the relevance of his account of competitive liberal desiring-machines for an understanding of relations among states. Hobbes offers an account of the way things should be given his claims about the assumptions that allow one to read the world as it is. He even gets away with claiming that we are free to the extent that we do what we must do. But he gets away with his account of political necessities through an act of persuasion, an act in which claims about reason, logic and science may play a crucial role but never quite obscure the extent to which they are simply claims. Hobbes may be without peer in making it seem as though one's secular duty and the will of the sovereign are consistent with geometrical reason, natural law or even divine will, but his skills lie in the art of making it seem, rather than in any firm grasp of the kind of philosophical foundations imagined by so many recent defenders of modern reason.[14]

So, in the first instance, notions of a crisis in international relations theory have to be disentangled from the overdetermined rhetorical pretence that we are confronted by a sudden move from foundations to a foundationless universe, from a world that believes (as it is so often said, without even a trace of irony) in Enlightenment and the rest to one that does not. It is much more useful to treat a significant part of those literatures that are coded as postmodern as, in the first instance, fairly simple attempts to remind ourselves of the conditions under which we have chosen to forget about the arbitrary acts of will, about the sovereign authorities that have in turn become the possibility conditions for values and institutions we have come to value so highly, or to treat with a renewed scepticism. This is why so many of the so-called critical theorists of international relations have developed a persistent interest in how state sovereignty works, how it constitutes and insinuates itself among a variety of other practices that have come to be taken for granted as the natural and rational condition of human affairs. While they may indeed seek to contribute to fermenting a crisis in some other sense, such theorists have been most successful in drawing attention to a crisis that is already there, that is the very possibility condition for a theory of international relations at all. While their most immediate influences may come from twentieth century philosophers of language and identity, contemporary accounts of how state sovereignty is constituted and reconstituted are not entirely at odds with what Hobbes and other icons of political and international relations theory were doing quite some time ago.

IR and the Fate of the Political 227

There are three key sites at which the theory of international relations frames a specific understanding of crisis:

- It frames a boundary between the normal and the abnormal defined in relation to domestic peace and international violence. Much of what has become known as 'political realism' or the 'anarchy problematic' is in this sense an account of what *must* happen when norms are suspended. Crisis is defined as the need for exclusion, as the need for exceptions to the general rule. In this context, international relations theory does the dirty work for political theory; it focuses on those moments of modern politics when domestic norms are transgressed, when knowledge and value legitimately collide with power, when the rights of citizen-subjects end at the frontiers of violence. The single most striking feature of modern political thought, in fact, is the degree to which the distinction between political theory and international relations theory, between political community within and the absence of political community without, has been naturalised and abandoned. The fear on the part of so many theorists of international relations induced by questions about political theory, or worse, political philosophy, is exceeded only by the almost universal fear on the part of political theorists of stepping beyond the limits of the singular political community, unless, of course, they pretend that 'the world' is simply a larger version of their idealised image of the state.

- Conversely, it frames a boundary between the normal and the abnormal defined in relation to a theory of developmental history. All the problems posed by a theory of progress in a multicultural and multi-jurisdictional world are resolved through a claim about universal participation in a universal set of norms institutionalised in the states system. The underlying assumption, then, is that those cultures and jurisdictions that have not yet fully achieved the norms of the modern states system can be encouraged to do so. The still powerful distinction between First and Third Worlds, for example, works this way. Crisis is framed here more as a need for inclusion, for bringing certain external domains into the norms of domesticity. It is in this respect especially that the theory of international relations has much in common with anthropological framings of the temporal relations between self and other.

– It also frames a boundary between present and future in relation to both the need for and the tragic impossibility of moving beyond the present state of crisis towards some more inclusive, or peaceful structure of humanity, global governance, and so on. The international system is constituted as a state of crisis, and there is no alternative to living in this state of crisis. The utopian moment in the theory of international relations is simply an affirmation that the world must remain as it is. The sense of crisis is thus doubled in that the mechanisms through which crisis in the international system has been managed are said (by some) to be no longer sustainable, but the very aspiration for an alternative only confirms the claim that alternatives are out of the question.

The traditions that have been identified as political realism and political idealism both participate in the articulation of this discourse of crisis. Political idealism is nothing more and nothing less than an attempted universalisation of an idealisation of a statist political community, of the legacy of the *polis* as the epitome of the good life. Political realism is nothing more and nothing less than an insistence on the limits, the point of crisis, in this idealisation. They are mutually constitutive positions and they continue to inform almost all other attempts to understand ways in which human affairs are being or might be reconstituted.

The theory of international relations, then, expresses a particular way of understanding what a crisis is. The real difficulty before us is that according to this understanding of crisis there can be no way out because we are so wedded to or dependent upon those things – the modern state, the modern subject, modern conceptions of space, time, identity, and community – that require a theory of international relations as the crucial limit on their possibility.

THE SUBSTANTIVE OR THE METAPHYSICAL?

Kal Holsti canvasses a wide range of arguments that have led many others to the conclusion that the theory of international relations is indeed in crisis. He also examines several alternatives that might plausibly sustain a Restoration more appropriate to contemporary circumstances. The tone of the analysis is distinctively and helpfully ambivalent, for all that it is framed as a provocative expression of concern about contemporary incoherences. A positive sense of the opportunities opened up by these controversies is not entirely smothered by the contrary sense that something is very seriously wrong. In articulating this (almost) ambivalent judgement, however, it seems to me that he both overstates and understates the seriousness of contemporary controversies.

On the one hand, Holsti overstates the fragmentation of the field. Intellectually, the discourses of political realism and political idealism are still alive and well. Hobbes still gets read as a version of Machiavelli and the opposite of Kant, and the categorical structures of the discipline follow suit. Much of the overt language of 'new' approaches, both 'mainstream' and 'critical', is but a thin crust spread upon old categories: the historical/practical constitution of norms is forced into the language of utilitarian rationality; romantic pluralism is blessed with the label of postmodernism in the manner of contemporary American cultural studies; Hegelian/Marxian/Lucacsian/Mannheimian struggles with claims about class consciousness are turned into rhetorics of identity politics. The world is still categorised and dissected as a hierarchy of supposed 'levels'. And in a more prosaically sociological context, it is certainly not clear to me that the practices of scholarly recruitment have been radically or even marginally overturned.

On the other hand, Holsti underestimates what is at stake in some of the moves that seek to resist established conventions. Let me take two of his central complaints: that 'theorists have *chosen* to focus on epistemological and scholastic problems rather than substance [and the absence of any] essential core problem to investigate.'[15]

The charge that too many theorists engage in abstract navel-gazing is common currency, used variously to spur researchers to simply do their research rather than just talk about how to do it, or to face up to the urgent demands of the day and quit whining about trivialities. And there is, it must be admitted, often something to this charge, though usually much less than it is made to seem. The charge usually accompanies a transparent defence of a severely restricted account of legitimate research strategies or (from my perspective at least) a very thin account of political responsibility and the role of the intellectual. It also tends to express a naive understanding of the politics of language, and to lapse into dualistic understandings of a language separated from a world to which it refers and complaints about a concern with mere 'text'.

The crucial difficulty here, however, is that questions about epistemology, language and ontology *are* substantive. This is one of the key lessons that theorists of international relations ought to be able to learn from reading the few chapters of *Leviathan* that Hobbes wrote before telling his story about a state of nature, let alone from the libraries devoted to this theme over the past century or so. Like the libraries devoted to the theory of the state (also, it must still be said, infrequently visited by theorists of international relations), many of those interested in 'scholastic' problems are concerned precisely with questions about authorship and legitimate authority, with the conditions under which claims and practices succeed and fail, with the shifting boundaries of power and knowledge, and with the construction of subjectivities, communities and exclusions. It is true that the jargons, references and levels of abstraction at work in some of these discussions can be quite obscure to the unitiated, sometimes unnecessarily so, but hardly more so than discussions of liberal trade theory, military technology, or even baseball and cricket.

The significance of Hobbes's discussion of language, and all that follows from it, on the other hand, is, I think, absolutely crucial. Without it the very idea of modern sovereignty is unthinkable. Who gets to speak? Who gets to speak authoritatively? How do those who claim to speak authoritatively get away with it? Such questions ought to be of some significance for debates about any crisis in international relations theory, both in relation to how 'the world' works and how 'the theory' works in relation to 'the world'.

This is also why the claim that we have no essential core problem to investigate is, I believe, seriously misleading. Holsti questions whether there is 'a commanding problem or challenge that animates theory and debate.'[16] It is true that there is now a much wider spread of research interests expressed in the literature, and there is, thankfully, and outside of a few well known university departments at least, no longer such an overbearing sense of hegemony on the part of any single theoretical tradition. Judged at this level, and keeping in mind the kaleidoscopic scenes portrayed by journalistic accounts of and informed commentaries on what is currently going on in the world, it is difficult to see how this cannot be anything other than a sign of progress. The single-minded obsessions that shaped the discipline during the Cold War managed to obscure a great deal of what could have been placed under the rubric of international relations. At least one might hope that similar obsessions do not manage to reduce analysis of the diversity of problems and patterns shared by the peoples of the planet to the narrow concerns of a few powerful societies.

But there is a sense in which for all the fragmentation of diverse research interests, the theory of international relations is powerfully shaped by a fairly small group of problems, and perhaps even by a single question, though one with many corollaries and variations. It is the simple question about human identity, about who we are and how we might live together whoever we are. This is the question that has been answered by claims about the modern subject. We are, supposedly, self-representing, self-developing, self-identical subjects. Or, in the form that is crucial to the theory of international relations, we are supposedly free and responsible citizens of sovereign states. Modern politics, and the modern theory of international relations, is grounded in the claim that this is a sufficient, and perhaps even necessary and inevitable answer to this question, a question that is certainly not a mere matter of philosophy and other abstract pursuits.

This question has not gone away. Claims about state sovereignty, however, are certainly looking a bit ragged around the edges. They have long been a bit ragged around the edges, of course, which is why debates about realism, idealism, war, intervention, the place of international law, organisation and ethics, and so on have been so persistent. But the real force of contemporary debates lies in the degree to which the early modern insistence that claims about citizenship have priority over all claims about humanity, and indeed that one can only achieve one's humanity by paradoxically submitting to the necessities of citizenship, is more than simply ragged around the edges. It is this insistence that is expressed by the principle, institutions and practices of state sovereignty.

If the theory of international relations is in crisis, it seems reasonable to expect that the crisis is somehow linked to the decreasing credibility of its constitutive categories and assumptions. It seems especially reasonable to suspect that the crisis is linked to the decreasing plausibility of the claim that state sovereignty is a sufficient answer to the question of who we are. This does not suggest the lack of any core problem. From matters of ethnicity and nationalism, to challenges from feminists and postcolonial theorists, to claims about global community and cosmopolitan democracy, to attempts to dislodge the congealed assumptions about identities and knowledge claims embedded in various 'texts', to imaginary projections of the future shape of global governance or a global civil society, to analyses of the rearticulation of (non-territorial) patterns of inclusion and exclusion in relation to emerging forms of political economy, to attempts to come to terms with inherited conceptions of Otherness in the framing of self-identities, to commentaries on the reshaping of communities and alliances in particular regions: the continuing plausibility or otherwise of the modern claim to a sovereign identity is not difficult to detect as a nagging, inchoate but irresistible problem that no theorist of international relations can ignore. Exactly what one makes of this question once one admits its significance is a more difficult matter. But I would interpret our present position in exactly the opposite way from Professor Holsti and note the increasing clarity with which problems posed by the fragility of claims about sovereignty as the answer to our most persistent questions drives the research agendas of quite diverse groups of scholars. Much of what has been construed as an anarchy loosed upon the cosy world of international relations as a discipline is more positively understood precisely as an attempt to come to terms with the limits of political possibility as we have come to know them, to love them and to hate them in the modern age.

REPRISE

That the theory of international relations is in some kind of crisis is hardly surprising, I believe, because the principle, institution and practice of state sovereignty is indeed in a condition of crisis, though what this crisis involves precisely remains unclear. But as a principle, institution and practice of international relations, state sovereignty is only one moment of a broader set of principles, institutions and practices of what might be called a culture or even a civilisation. There is absolutely no point in pretending that it is possible to speak about state sovereignty without some understanding of how it expresses, and works in relation to, the constitutive principles of modernity. 'Accept my assumptions,' said Hobbes, 'and one can come to no other conclusion.' His assumptions, of course, were precisely about language, memory, dreams, reason, passion, God, angels, atoms, geometry, freedom, necessity, finitude and the logical impossibility of modern individualism. And there is no point in dismissing Hobbes because he is a long dead philosopher, or even a dead white male, because Hobbes's assumptions are alive and well in almost all of us. These assumptions, or something very close to them, have been internalised as conditions under which we have become able to understand ourselves for what we are. This is what we need to become much more self-conscious about.

And this, it seems to me, is what in large part informs so many otherwise disparate attempts to reconstitute the theory of international relations on some other basis. It is this, also, that leads me to be more sceptical about some of these attempts than others: about, say, attempts to mediate between cosmopolitan and communitarian, or communitarian and individualistic forms of liberalism; or to revive Aristotle, Kant or Hegel as model synthesisers; or to discover some ethics on which to ground a politics; or even to recover a more historically sociological or economic understanding of the modern state and states system. It leads me to be especially sceptical of any attempt to pose questions about future possibilities in terms of an aspiration for universality, as a revamping of Kantian or Hegelian accounts of a cosmopolitan community of some kind. While undoubtedly signs of serious intellectual life, to my mind these are all moves that whatever their other (and often many) virtues, are in danger of reproducing the categories that keep us more or less where we are.

A large part of the difficulty in coming to terms with debates about state sovereignty is that its character and genealogy are so often framed monologically, undialectically. The notion of a monopoly in a specific territory prevails. Accounts of what state sovereignty is are written in terms of an act of fragmentation, of particularisation. But state sovereignty is a binary or dialectical not a monological category. Our most important silences have not been about things that go on at the edges of modern life, in the realm of international relations, but about those that go on right at the centre. The conventions of modern political theory are in even greater trouble than the conventions of international relations theory. The presence of a sovereignty here requires the absence of a sovereignty there. The sovereignty of any particular state requires a system that allows that particular state to exist as a sovereign jurisdiction. Sovereignty is not properly understood as an expression of an extreme – as constitutive of anarchy without, or of order within – but precisely as a middle ground, as a point of intersection between extremes: between inside and outside and between universality and particularity. It is in fact, an expression of the characteristic manner in which the modern world has been able to answer questions about the relationship between one and many, between the multiplicity of things and people, on the one hand, and the unity of the world and humanity on the other, through a spatial differentiation between inside and outside. This is why it must be understood as an answer to a question – about who we are as particular participants in something much grander – before it can be understood as an institution and practice that can be examined on its own terms. The crucial puzzle before us is whether it remains a satisfactory answer to this question. Many literatures now make a persuasive case that it is not.

As a corollary of this, another part of the difficulty in coming to terms with debates about state sovereignty is the continuing resort to the discourse of presence and absence that state sovereignty itself serves to put into effect. Questions about the continuing presence or imminent absence of state sovereignty are especially misleading because they simply reaffirm state sovereignty as a concept of extremes, not of reconciliation. It is the capacity of state sovereignty to mediate between the claims of universality and those of particularity that is in question, not some shift from a world of particularities to one of unities. Again, bluntly, the early-modern trade-off between men and citizens, between understanding that one can become human only by becoming a citizen of a particular state, or in later terms, of a particular nation, is no longer as persuasive as it has been, though it is still undoubtedly more persuasive in some places than in others. This is in large part why the political cartographies of the contemporary world seem so messy. It is simply more and more difficult to make sense of who we are or how things work by assuming that all relations between unity and diversity can be resolved on the spatial terrain of the modern state.

Substantively, this is where the theory of international relations can most usefully be said to be in crisis. What we can see in the disarticulation of the theoretical enterprise, I believe, is an expression of the disarticulation of relations between unity and diversity that the modern project so successfully fixed on the spatial terrain of the sovereign state and the sovereign subject more generally. But such a disarticulation should not be conceived as the kind of crisis expressed as the limit of modern subjectivities. We are not being pushed to the edge of a dangerous relativism or a dangerous anarchy. We are, perhaps, all becoming more and more aware of the dangerous edges that have sustained our most cherished political aspirations since the early-modern era. There is certainly a danger that scholars will keep trying to act like sovereigns of old and try to police the boundaries between authentic community and scholarship within, and supposedly dangerous relativisms without. But it would be more appropriate, I believe, if we all become more modest and recognise the difficulties we will all encounter in trying to make sense of who we are and what is going on in the world given that we have only the most fragile and fleeting idea of what it might now mean to speak of the relation between universalities and particularities on terms other than those required by the discourses of sovereign subjectivity. Either the theory of international relations will take up take such apparently abstract or philosophical matters or it will cease to have any interest at all for those concerned to understand the practices of contemporary world politics.

NOTES

1. Some aspects of this argument are explored in R.B.J. Walker, *Inside/Outside: International Relations as Political Theory* (Cambridge: Cambridge University Press, 1993); and Walker, 'International Relations and the Concept of the Political,' in Ken Booth and Steve Smith (eds.), *International Relations Theory Today* (Cambridge: Polity Press, 1995), pp. 306–27.
2. More explicitly philosophical attempts to treat politics as a troubled category include Dana Villa, *Arendt and Heidegger: The Fate of the Political* (Princeton, NJ: Princeton University Press, 1996); Simon Critchley, *The Ethics of Deconstruction: Derrida and Levinas* (Oxford: Blackwell, 1992); and Richard Beardsworth, *Derrida and the Political* (London: Routledge, 1996).
3. One need only browse the past few volumes of *Millennium: Journal of International Studies* to get a sense of what is going on, though I also have a personal stake in seeing *Alternatives* as an especially significant vehicle for theoretical innovation.
4. As this has been posed recently by William Wallace, among others. William Wallace, 'Truth and power, monks and technocrats: theory and practice in international relations,' *Review of International Studies* (Vol. 22, No. 3, July 1996), pp. 301–21. Quite apart from what seem to me to be some particularly willful misreadings of texts and theoretical traditions, Wallace's analysis is especially interesting for (i) its replication of a simple distinction between ideas and reality that ought to have been killed off by Marx, Weber and Mannheim, among many others, and (ii) its happy equation of practical relevance, or the responsibilities of the 'intellectual class in modern society,' with an attention to government policy. It is perhaps unfortunate that questions about truth and power are not quite so straightforward, but it is certainly even more unfortunate that influential scholars would still have us believe that they are.
5. I recognise that a focus on supposedly capitalist or supposedly gendered rather than supposedly modern accounts of political identity would lead to significant differences in emphasis – but only of emphasis – in what follows.
6. This confusion is especially pervasive in recent British accounts of a Critical Theory of international relations that take their cue from certain readings of modernity as an incomplete project associated with Jürgen Habermas. In this context, especially, foundationalist claims about ethics provide a convenient but unpersuasive way of avoiding an engagement with the constitutive principles of modern politics.
7. For a helpful attempt to make sense of recent competing claims about globalisation, see Jonathan Perraton, David Goldblatt, David Held and Anthony McGrew, 'The Globalization of Economic Activity', *New Political*

Economy (Vol. 2, No. 2, July 1997). Examples of recent literatures on this theme include Roland Robertson, *Globalization: Social Theory and Global Culture* (London: Sage, 1992); Mike Featherstone (ed.), *Global Culture: Nationalism, Globalization and Modernity* (London: Sage, 1990); Mike Featherstone, Scott Lash and Roland Robertson (eds.), *Global Modernities* (London: Sage, 1995); Paul Hirst and Grahame Thompson, *Globalization in Question* (Cambridge: Polity Press, 1996); Martin Albrow, *The Global Age* (Cambridge: Polity Press, 1996); Yoshikazu Sakamoto (ed.), *Global Transformation: Challenges to the State System* (Tokyo: United Nations University Press, 1994); and Richard Falk, *On Humane Governance: Toward a New Global Politics* (Cambridge: Polity Press, 1995).

8. K. J. Holsti, 'Along the Road of International Theory in the Next Millennium: Three Travelogues,' paper prepared for the joint Japan Association of International Studies – International Studies Association Conference, Mukuhari, Japan, September 20–22, 1996.

9. Holsti, *ibid.*, p. 11.

10. Holsti, *ibid.*, p. 28.

11. Holsti, *ibid.*, p. 1.

12. Not least in relation to the late-medieval/early modern metaphysics of time and eternity or finitude and infinity, though this is not a theme I intend to take up here.

13. One of the few points made by William Wallace in his recent comments on related themes which bear much scrutiny, and it is an important point, concerns his observation that many recent forms of critical theory have been in play for a long time and that it is unhelpful to exaggerate their novelty. (Wallace, *op. cit.*, pp. 309–11) The forms of logical positivism that were so influential in the construction of modern American social science were effectively part of a counter-revolution against the multiple forms of neo-Kantian, Nietzschean, Bergsonian and many other historicisms that sprung up in the late-nineteenth century and reappeared, though considerably reshaped by other later influences, in both Frankfurtian and Parisian forms of critical theory. Many of the issues at stake in the earlier period still resonate in some of the early texts of some of the classical realists, mainly in forms derived from Weber and Schmitt (Morgenthau) or Weber, Lukacs and Mannheim (Carr), though the resonances did not survive for very long. Quite a lot of intellectual history, much of it strenuously contested, would, of course, need to be invoked in order to make much sense of the almost complete surrender of even the emigre scholars now identified as classical realists to the peculiar mix of nationalist *machtpolitik* and pre-Kantian empiricism that has sold so many bizarre textbooks and launched the careers of so many policy advisors on national security.

14. For a highly detailed historical analysis of Hobbes's rhetorical strategies, see Quentin Skinner, *Reason and Rhetoric in the Philosophy of Hobbes* (Cambridge: Cambridge University Press, 1996).
15. Holsti, *op. cit.*, p. 25 and *passim*.
16. Holsti, *ibid.*, p. 7.

Index

accountability, 6, 133, 154, 163, 175–6, 186, 195–6
agency, 129, 151, 156, 158–9, 163, 173, 176, 178
agency/structure, 151, 152,154, 156, 158–9
agonistic political argument, (*see politics*)
Alexandrowicz, Charles, 36
anarchical society, 23, 28, 37–8, 40
anarchy, 4, 54, 178, 214, 227,232, 234–5
Anderson, Benedict, 201
Annales School, 55
anti-foundationalism, 129, 134, 192
Arendt, Hannah, 131, 137
Aristotle, 56, 72–3, 197, 233
Austin, J.L., 78
authority, 13, 17, 21, 35, 101, 107–8, 153, 162, 168, 176, 180n.17, 187–8, 196, 203, 207, 215–16, 225, 230

balance of power, 17, 36
Barrington-Moore, J.B., 57
Benhabib, Seyla, 129, 131, 134
Brilmayer, Lea, 188
British Committee (for the Theory of International Politics), 34, 36
Brown, Chris, 26, 179n.4, 181n.24
Brubaker, Rogers, 198
Bull, Hedley, 5, 6, 9, 18, 19, 27–34, 36–40
Burke, Edmund, 56
Burton, John, 168

Butler, Judith, 130, 132–4, 147n.9
Butterfield, Herbert, 34–5

capitalism, 98, 99, 100–2, 104, 107–8, 110, 112–14, 116, 126, 153, 161, 168, 172–3, 176, 180n.4, 214, 236n.5
Castells, Manuel and Peter Hall, 102
Cavanagh, John, 112
Chase-Dunn, Christopher, 112
citizenship, 39, 84, 155, 161,170, 172, 184n.50, 185–7, 198–9, 232
civil society, 100, 109–11, 113, 115–16, 198
  global, 18, 101, 185, 232
  international, 63, 77
civilisation, 53–5, 80, 82, 85–7, 91, 156, 169, 233
Cold War, 49, 50–1, 86, 88–9, 212, 216, 221, 224, 231
colonialism, 61, 62, 65–6
comparative contingency, 59–67
communicative action, theory of, 79, 125, 129, 141, 150n.44
communitarianism, 172, 181n.24, 233
community, 2, 24, 27, 29, 31,35, 38, 47, 52–3, 69n.12, 69n.13, 99, 111, 114, 121, 153, 154, 156, 160, 163, 169, 171–4, 176, 181n.26, 187, 191–2, 195, 197, 199, 200–1, 204–5, 216, 227–9, 232, 235
conceptual tools, 8, 12–15, 142, 160, 185
Connolly, William, 1–4, 7, 9,17,

20, 26
consent, 12, 60, 63, 110,
  186–92, 196–7, 199, 205–6
cosmopolitanism, 30, 51, 73,169,
  181n.24, 201, 208, 233
Cox, Robert, 112, 140, 142
Crick, Bernard, 1–3, 5, 10,72–3
critical theory, 8, 11, 60, 123–5,
  127, 129, 130, 132–6, 139–40,
  142, 146n.5, 222, 236n.6,
  237n.13
culture, 10–11, 14, 47–71,72–97,
  99, 101, 104, 106–7, 114, 125,
  130, 132, 136, 138, 139,
  141–3, 148n.31, 150n.45, 160,
  163, 167, 169, 177, 180n.18,
  187, 207, 212–5, 220–1,
  223–4, 227, 233
cultural imperialism, 48, 54,60,
  62, 145

Dahrendorf, Ralf, 99
democracy, 5, 20, 49–50, 76–7,
  86, 93n26, 104–5, 115, 152,
  154, 159, 164, 167, 176,
  177–8, 181n.23, 193, 195, 201
  cosmopolitan, 75–7, 232
  liberal, 72, 76–7, 82–3,
  86–90
Deng Xaoping, 75, 82, 87,93n.31
Derrida, Jacques, 52
Descartes, Rene, 220, 225
difference, 18, 48, 52, 53, 61,106,
  134–5, 138, 142–4, 165, 175,
  184n.52, 202
Dilthey, W., 78
domestic, the, 5–6, 12, 14, 66,83,
  116, 186, 217, 227
  domestic/foreign divide,
  151–84
Drainville, Andre, 112
Dworkin, Ronald, 195

Durkheim, Emile, 204

emancipatory politics, 40, 61,64,
  111, 123, 125–8, 130, 136,
  141, 146n.5, 151, 217
English School, 28, 34–8, 55,
  152–3
Enlightenment, 56, 126, 133,202,
  224–6
epistemology, 67, 98–9, 106, 115,
  123, 127, 129, 130–1, 141,
  216, 222, 230
essentially contested concept,
  1, 3, 4, 7, 9, 19–23, 26, 28, 34,
  37–40, 43n.37
ethics, 59, 79, 127, 154–5, 157,
  180n.18, 212–14, 231, 233,
  236n.6
ethnicity, 48–51, 62, 66, 82,163,
  177, 202–4, 210n.29, 232
ethnocentrism, 74, 86, 96n.79
Etzioni, Amitai, 69n.13, 172

Falk, Richard, 112
Fanon, Frantz, 65
feminism, 123–50, 223, 232
Flax, Jane, 131
foreign, the, 12, 108, 188, 197
foreign policy, 12, 31, 107,
  150n.44, 150n.45, 151–84
  foreign policy analysis, 5, 151,
  153, 176, 179n.4
foundationalism, 128–32, 134,
  139, 148n.22, 192, 225
fragmentation, 214, 234
Frankfurt School, 71n.31, 129,
  146n.5, 237n.13
Fraser, Nancy, 124, 134–45,
  148n.31, 148n.33, 149n.39
Fukuyama, Francis, 63, 97n.85
(neo)functionalism, 152, 159,165,
  168, 174, 176

Index

Gadamer, H.G., 78
Galbraith, John Kenneth, 98
Gallie, W.B., 3, 20, 25, 43n.37
Galtung, Johan, 52
Gellner, Ernest, 202
global governance, 75–6, 155, 217, 228, 232
globalisation, 11, 48, 51, 53, 55, 65, 72–3, 76–7, 82–91, 96n.79, 98–122, 151, 158, 166–8, 175–6, 185, 201, 214–16, 223
glocalisation, 83–7, 90
Gottingen School (German Historical School), 36
government, 3, 28, 73–4, 99, 104, 106, 113, 149n.33, 155, 157–8, 163–5, 168–70, 172–3, 176, 178, 186, 188–90, 193–4, 198, 220
Gramsci, Antonio, 10, 47, 59–67, 99, 110, 116, 146n.5
Grotius, Hugo, 25, 30–2, 34–5, 37, 40
group, the, 12, 161, 181n.24, 187, 194, 197–8, 206–7

Habermas, Jurgen, 78–80, 85,88, 125, 129, 131, 146n.5, 150n.44
Harding, Sandra, 133, 142, 145
Hart, H.L.A., 21
Hartsock, Nancy, 129–31
Hegel, W.F., 229, 233
hegemony, 47, 60, 61–5,70n.29, 74, 76–7, 82, 109–10, 116, 137, 180n.14, 231
Held, David, 68n.2, 75–7, 89, 93n.31, 185
Herder, J.G., 53, 203
hermeneutics, 78
Herz, John, 153, 176
Herzog, Don, 190
Hirst, Paul and Grahame

Thompson, 74–5, 77, 89
Hobbes, Thomas, 29, 32, 189, 194–6, 202, 213–14, 220, 225–6, 229–30, 233
Hoffman, Mark, 124
Hoffmann, Stanley, 223
Hollis, Michael, 51
Holsti, Kal, 212–13, 216–19, 229, 231–2
human rights, 38, 75–6, 82, 84, 155, 157, 172, 177, 183n.45, 185, 201, 206–7
Hume, David, 192, 201
Huntington, Samuel, 49, 54–6, 62, 71n.34, 76, 85–6

identity, 18, 48, 51, 53, 56–7, 59, 61–2, 64, 78, 84–5, 87–8, 126, 128–31, 134–5, 138, 142–3, 151, 154, 167, 169–75, 181n.26, 198–9, 202–4, 207, 211n.42, 214, 217, 221, 223, 226, 229, 231, 232, 236n.5
ideology, 65, 71n.31, 100–1, 106, 116, 145, 208
interdependence, 54–5, 73, 77, 105, 156, 164, 175, 185, 223
international community, 24, 26, 171
international law, 17, 23–5, 43n.37, 94n.43, 150n.45, 157, 231
international order, 27, 38, 73
international political economy, 5, 60, 85, 100, 107, 115, 136, 138, 143, 151, 168, 214, 216–17, 232
international public sphere, 140
international relations theory, 13, 17–19, 34, 37, 40, 50–2, 123, 213–16, 218–27, 230, 234

international society, 2, 9, 17–19, 24–40, 62–3, 72, 74, 78, 80–4, 86, 89–90, 153
international system, 5, 26–7, 29, 49, 52, 56, 59, 62–4, 67, 98, 152, 169, 177, 186, 228
Islam, 49, 58, 65, 85, 86, 206

James, Alan, 27
Jessop, Bob, 102
Johnson, Lyndon, 166–7
justice, 5, 28, 30, 40, 90, 111, 115, 136, 138, 166, 192, 194, 203, 206

Kahler, Miles, 177
Kant, Immanuel, 56, 77, 222, 225, 229, 233, 237n.13
Keohane, Robert, 51, 54, 153, 156
knowledge/power, 40, 125, 127, 130–1, 141, 149n.37, 220, 227, 230
Kuhn, Thomas, 224

Lapid, Yosef, 185
legitimacy, 12, 24, 65, 135, 152, 176, 185–90, 192, 198, 202, 205, 222
Lennon, Kathleen and Margaret Whitford, 133
(neo)liberalism, 51, 53–5, 72, 75, 84, 86, 98–9, 101, 104, 107, 109, 111–16, 152, 159, 172, 180n.14, 186–7, 191–3, 200, 205, 214, 221, 226, 230, 233
liberal paradox, the, 72, 74, 76–7, 89
Linklater, Andrew, 39, 139, 142, 150n.44
local, the, 11, 14, 48, 62, 77, 84–6, 88–9, 107, 111–16, 128, 158–9, 163, 174, 180n.15, 207, 216
Locke, John, 24, 35, 188–91, 198
Lukes, Steven, 20

Machiavelli, N., 31, 32, 60, 220, 222, 229
MacIntyre, Alasdair, 22
MacKinnon, Donald, 34–5
Madison, James, 193–4
Mao Zedung, 81, 88
Mayall, James, 27
Marxism, 5, 52–4, 81–2, 129, 136, 146n.5, 152–3, 201, 229, 236n.4
Mason, Andrew, 20
media, 48, 50, 53, 66, 83, 106–7, 149n.33, 162, 177, 180n.17
membership (*see also* citizenship), 12, 24, 26, 30, 32–3, 187, 191–3, 197–8
metatheory, 128, 134–5, 139–40, 142, 145
Miller, David, 22
modernity, 40, 85–7, 91, 103, 173, 192, 200, 201–6, 208, 210n.39, 215, 225, 233, 236n.6
moral, 2, 18, 20, 21, 26–7, 31, 38, 51–2, 54, 57, 59, 66, 69n.12, 70n.29, 83, 87, 90, 129, 131, 142, 150n.44, 163, 172, 181n.24, 183n.45, 192, 201, 203–04
Mouffe, Chantal, 17
Mulgan, Geoff, 169

Nardin, Terry, 90
national interest, 31, 160
nationalism, 48–51, 54, 57–9, 61–2, 64–6, 71n.36, 75, 77, 80–8, 160, 171–3, 185, 201–5, 210n.39, 212, 232

Index

natural law, 34, 80, 192, 206, 226
Nietzsche, F., 222, 237n.13
non-state actors, 18, 26, 32
non-governmental organisations (NGOs), 75, 111, 140, 158, 164
normative theory, 5, 49, 51–3, 66, 132–3, 136, 151, 155, 170, 176, 178, 188, 217, 219
norms, 63, 203, 227–9
Nye, Joseph, 51, 54, 63, 153, 156

obligation, 24, 26, 31, 34–5, 39, 181n.24, 188–200
Offe, Claus and Ulrich Preuss, 202
ontology, 126, 129, 140, 216, 221, 225, 230
order, 4, 14, 17–18, 26–7, 32, 38, 61, 125, 141, 191–2, 194–5, 198, 202–3, 222, 234

particularism, 13, 53, 141, 208, 234–5
peace, 27, 49, 51, 56, 86, 90, 159, 197, 222, 227–8
Plato, 72, 197, 199–202
pluralism, 86, 229
Polanyi, Karl, 108, 110
politics/the political (*see also authority, community, ethnicity, feminism, identity, obligation, power, state, society, values*)
as activity (practices), 2–3, 6–11, 14–15, 26, 30, 39–40, 57, 60, 72–4, 76, 89–91, 100, 109, 115, 123, 126–7, 131–3, 139, 141, 143–5, 146n.5, 155–8, 175–6, 184n.50, 185, 187, 190, 200, 207, 212, 214–16, 235
as actor, 5, 11, 74, 123, 125, 157, 163, 191, 192
agonistic political argument, 8–10, 14–15, 17, 19–23, 38–41, 159
political culture, 72, 74, 83, 87, 93n.26, 177
as discourse and contest, 1–11, 13–15, 17, 21–2, 30, 58–9, 64, 72–3, 76, 78, 82–3, 91, 127–8, 130–1, 135–9, 141, 143, 149n.37, 150n.45, 151, 154, 156, 158, 164, 186, 193, 201–2, 204, 206–7, 212–13, 215, 217, 221, 228–30, 234–5
institutions, 3, 5, 29, 34, 39–40, 48, 54, 82, 89–91, 103, 106, 168, 176–7, 191, 202, 216
meanings of, 1–2, 4–8, 17–19, 22, 25, 35, 47, 53, 72–3, 75, 79, 89–91, 103–5, 109, 111, 127–9, 131–40, 142–4, 148n.31, 148n.33, 150n.44, 153, 155–9, 173–4, 178, 180n.18, 186–7, 192, 197, 200–1, 203–5, 207–8, 212–16, 218–19, 223, 225, 233, 236n.6
official/discursive political, 136–139, 148n.33
system, 54, 60–1, 65, 98, 192, 194
theorisation of, 2, 4, 11, 17–20, 22, 35, 49, 51, 52, 55–6, 124, 127–9, 131, 135–6, 138, 140–5, 159–60, 164–5, 176, 186, 191–2, 199, 201, 205–6, 213–15, 219–20, 227
world, 9–13, 18–19, 23, 25–7, 29, 32–4, 39–40, 72–97, 98–122, 156, 160, 167, 175, 214, 216
positivism, 18, 25, 30, 31, 123–4,

176, 212, 222, 237n.13
post-positivism, 8, 11, 123–4, 134, 139, 140–44, 146n.4, 222
postmodernism, 11, 52, 79, 99, 102–3, 123–35, 138–0, 142–3, 145, 149n.39, 154, 166, 170, 172, 176, 215–16, 225–6, 229
poststructuralism, 1, 4, 8, 135, 154, 221
power, 2–5, 10, 14, 31, 47, 49–53, 58–67, 72–4, 79–80, 82, 100, 104, 106–10, 125, 127–8, 130, 133, 139, 141, 143, 149n.33, 150n.44, 155–7, 159, 162, 168–9, 176, 183n.41, 186, 189–91, 196–7, 207, 220, 227, 230, 236n.4
public opinion, 166, 170, 195

rationalism, 23, 31, 35, 85, 86
Rawls, John, 191–2
realism, 1, 4, 8, 12, 23, 31–2, 39, 49–50, 54, 124, 152–4, 156–7, 159–60, 166, 178, 179n.4, 185, 214, 219, 227–9, 231, 237n.13
reason, 24, 85, 125–8, 214, 222, 224–6, 233
reflectivism, 152
regime theory, 55
regionalism, 105, 107, 111–12, 114–16, 165, 167, 173–5, 186
regionalisation, 100, 110, 151
Rengger, Nicholas, 124,
representation, 12, 131, 187–8, 191–7, 199–200, 203–5
resistance, 8, 11–12, 77, 80, 100, 110–16, 130, 132, 142–4, 207
revolutionism, 23–5, 28, 31–2, 39, 43n.37, 156
rights, 111, 115, 121n.85, 138, 143, 189, 196, 198, 201, 206–7, 227

civil, 76
individual, 32, 35
natural, 35
political, 76, 193
Robertson, Roland, 84
Rosenau, James, 51, 184n.51
Rousseau, Jean–Jacques, 56, 189
rules, 17–18, 24, 26, 29, 31–4, 36–7, 39, 60–3, 78, 195, 198

Schmitt, Carl, 197, 225, 237n.13
Scholte, Jan-Aart, 155, 159
security, 4, 50, 63, 77, 156, 159–60, 164, 169, 170, 237n.13
Sklair, Leslie, 113
Smith, Anthony, 207
Smith, Steve, 51, 146n.4, 157, 180n.18
social, the, 11, 17, 20, 48, 52–5, 57–8, 60–6, 70n.25, 70n.29, 71n.34, 74, 76, 86, 100–1, 105–6, 109–12, 114–15, 125–6, 128, 131, 135–42, 145, 150n.46, 159, 161–2, 166, 169, 171–2, 177, 202, 206–7
social agency, 8, 11, 89, 103, 106, 110, 125
movements, 5, 8, 11, 14, 55, 75, 100, 103, 107, 111–15, 116, 121n.85, 123–5, 133, 136–41, 143, 145, 149n.33, 150n.45
theory, 50–3, 124, 133–6, 139, 142, 145, 146n.5, 219
society, 3, 6, 9,12, 19, 25, 27, 32, 39, 55, 57, 61–3, 65, 70n.25, 70n.29, 71n.31, 73, 86, 100, 105–7, 109–11, 113, 116, 125, 127, 137, 151, 153, 159–64, 166–7, 170, 172–3, 175–8, 181n.26, 195, 206–7, 236n.4

society of states,
  (*see international society*)
Socrates, 199
solidarism, 30, 32
sovereign, the, 189, 194, 196, 203, 215, 225–6, 234–5
sovereign subjectivity, 13, 128, 214, 220, 235
sovereignty, 13, 17, 29, 35, 37, 40, 48, 58, 72, 77, 87, 106, 154, 156, 185–8, 204, 213–14, 225, 230, 232, 234
  state sovereignty, 13, 18, 39, 62, 75, 77, 107, 143, 153–4, 160, 162, 175, 187–8, 212, 214–15, 220, 223–6, 231–5
state, the, 2–3, 5–6, 10–15, 23–4, 26–30, 32–4, 36, 48–56, 58–9, 61–3, 65–7, 72–7, 80–3, 85–7, 89–91, 93n.31, 98–101, 103–04, 106–111, 113–14, 116, 136–7, 140–2, 149n.33, 150n.44, 150n.45, 150n.46, 151–65, 168–73, 175–7, 180n.17, 181n.23, 181n.26, 183n.41, 185–8, 194, 198–9, 201, 206–7, 212–21, 223–35
states-system, 18, 32–3, 36–7, 77, 110, 155, 168, 180n.12, 206, 213, 219, 220, 227–8, 233
Strange, Susan, 156, 168, 177, 180n.17
  structuralism, 4, 52–4, 101, 103–4, 124, 168
structures, 54, 58, 62–4, 70n.25, 105–6, 125, 128, 134–6, 139, 157, 161, 177
subject, the, 12, 24, 126, 129–31, 135, 140, 186–7, 190, 196
Sylvester, Christine, 128, 134, 144–5

technological change, 51, 63, 86, 98–100, 102–4, 107, 151, 166, 168, 170, 173, 203, 216
Toynbee, Arnold, 55
transnational, 48, 51, 66, 75, 99, 101, 106–7, 153, 155, 160, 162, 164, 168, 176–7
  corporation (TNC), 107, 141, 168
transnationalism, 53, 65, 157, 176
transnationalisation, 106, 109
Tuck, Richard, 35

universalism, 13, 52, 58, 61, 64, 74–5, 79, 85, 126–8, 131–2, 135, 140–1, 150n.44, 192, 201, 206–7, 227–8, 233–5

values, 20, 27, 30–2, 35, 53, 55–7, 59–4, 71n.31, 76, 115, 130, 157, 164, 167, 171, 226
van Heeren, Alexander, 36
Vernon, Raymond, 153
Vietnam, 151, 160–1, 165–7, 182n.34
Vincent, R. J., 32, 38, 83

Waever, Ole, 18, 146n.4
Wagar, W., 113
Walker, R.B.J., 113, 128, 139, 142, 154, 179n.8
Wallerstein, Immanuel, 52
Waltz, Kenneth, 54
war, 1, 4, 52, 159, 217, 231
Watson, Adam, 34
Weber, Cynthia, 153–4
Weber, Max, 1–4, 51, 58, 85, 88, 192, 222, 236n.4, 237n.13
Wendt, Alexander, 51, 54
West, the, 27, 30, 35, 48–50, 60, 62–3, 76–9, 84–5, 133, 141,

197, 207, 222, 224
Westphalia, 36, 39, 77
Wight, Martin, 18, 23–5, 27, 30–2, 34, 38, 43n.37, 156
Williams, Raymond, 53
women's movement, 111, 114, 129, 130, 132–3, 136, 138

Young, Robert, 90
Yugoslavia, former, 62, 156, 162, 171, 177

Zalewski, Marysia, 134
Zysman, John, 108